FOREWORD

The papers included in this Proceedings are prepared for a technical session, entitled "New Analysis Techniques for Structural Masonry," sponsored by the Committee on Methods of Analysis of the Structural Division, and were presented at the ASCE Structural Engineering Congress '85 held at Hyatt Regency Hotel, Chicago, Illinois, on September 16–18, 1985. The motivation to organize this session and to print this Proceedings originated with the Editor of this publication because of his membership on the Committee on Methods of Analysis of the Stuctural Division, and his involvement in research with regard to the behavior of composite masonry during the last few years.

Although masonry structures have been constructed throughout the world for thousands of years, most of this construction has been carried out by trial-and-error without necessarily having an indepth understanding of the behavior of masonry subjected to loads. The interest in masonry construction in industrialized countries decreased from the turn of the century to very recently due to the availability and popularity of newer materials, like steel and concrete, which are more versatile and have proven to be economical for many types of constructions. However, a new interest in masonry construction seems to have developed during the last twenty years as more economical and higher quality clay bricks, concrete blocks, and mortar have become available. In addition, with the use of the steel reinforced masonry, it has become possible to utilize load-bearing slender masonry construction for buildings as high as 15 stories. A parallel surge of interest in masonry research has also occurred in various countries which is evident from the three U.S. and Canadian masonry conferences, and two international masonry conferences, in addition to those that have been held recently in Eastern and Western Europe and Asia.

As no sessions in masonry construction, analysis and design, or research have been organized by any committee of the ASCE at its annual conventions or structural specialty conferences recently, the present session is intended to generate interest among the structural engineering practitioners and researchers for a better understanding of some aspects of masonry behavior. The authors participating in this session have been engaged in masonry research for many years, and some of them are known world-wide for their expertise on the subject. Though much of the current masonry research being conducted is experimental, better analytical methodologies are also being developed because of the availability of more sophisticated computational tools. This Proceedings and the present technical session focuses on the use of computational techniques that are currently being used in the analysis of masonry structures in Australia, Canada, Switzerland,- W. Germany, Scotland and the United States. It is hoped that these papers will be useful to engineers engaged in masonry design and will serve as a stimulant to many others to engage in masonry research for a better understanding of the masonry behavior.

Each of the papers included in the Proceedings has been accepted for publication by the Proceedings Editor. All papers are eligible for discussion in the Journal of Structural Engineering. In addition, all papers are eligible for ASCE awards.

As the Editor of this Proceedings, I wish to take this opportunity to thank each and every one of the authors for taking time off their busy schedule to write the paper. Because of the time it takes to have mail delivered to various parts of the world, it has not always been possible to provide sufficient time to various authors in the preparation of their manuscripts. I appreciate their understanding and cooperation in this regard, and still delivering the papers on time. Lastly, on behalf of the Committee on Methods of Analysis of the Structural Division, I wish to thank Ms. Shiela Menaker, Manager, Book Production, ASCE for her cooperation and help in the production of this Proceedings.

Subhash C. Anand

CONTENTS

An Inplane Finite Element Model for Brick Masonry
  A. W. Page, P. W. Kleeman, and M. Dhanasekar 1
Macroscopic Finite Element Model for Masonry Walls
  A. S. Essawy, R. G. Drysdale, and F. A. Mirza 19
Behavior of Cavity Masonry Walls Under Out-of-Plane Lateral Loading:
An Analytical Approach
  A. A. Hamid and F. A. Carruolo 46
Shear Design of Masonry Walls
  H. R. Ganz and B. Thürlimann 56
Load-Bearing Capacity of Masonry Walls in a Critical Buckling Condition Taking
Into Account the Restraining Effect of Floor Slabs
  W. Mann and E. Leicher 71
Wall / Floor-Slab Interaction in Brickwork Structures
  A. W. Hendry 87
Shear Stresses in Composite Masonry Walls
  S. C. Anand 106

Subject Index 129

Author Index 130

AN IN-PLANE FINITE ELEMENT MODEL FOR BRICK MASONRY

Adrian W. Page[1], Peter W. Kleeman[2], Manicka Dhanasekar[3]

SUMMARY

A finite element model for brick masonry is reported. The model takes account of non linear deformations and progressive local failure. The failure may occur in the joints alone or in a mode involving both bricks and joints. Macroscopic deformation relations and failure criteria have been derived from the results of a large number of biaxial tests on half scale brick masonry panels in which the orientations of the bed joint planes to the edges of the panel were varied. The iterative plane stress finite element program is based on eight noded isoparametric elements and is used to simulate the incremental loading and progressive failure of brick masonry under in-plane loads. The effectiveness of the program is demonstrated by comparing the computed behaviour with the experimental results from racking tests on five steel frames with brick masonry infills.

1. INTRODUCTION

The application of the finite element method to the analysis of brick masonry structures requires an appropriate material model. To fully define the behaviour of brick masonry, a large number of tests under a range of biaxial stress states need to be performed. This poses considerable experimental difficulties both in the complexity of the testing rig and the number of tests required. Most previous analyses of brick masonry have used material models based on average properties, with the influence of the mortar joints acting as planes of weakness being ignored. These simplifying assumptions were used because of the lack of more detailed information on material behaviour. Typical of these analyses were those of Samarasinghe et al (1981) and Ganju (1977). In an attempt to overcome these difficulties, Page (1978) proposed a finite element model which considered bricks and joints separately, with the joints possessing non linear deformation characteristics and only limited capacity in shear and tension. However, this model was incapable of predicting brick failure and because of the large number of separate brick and mortar elements needed, the analysis of large wall panels posed computational problems.

This paper describes a finite element model which incorporates realistic material characteristics derived from a large number of biaxial

[1]Senior Lecturer, Dept. of Civil Engg. and Surv., Univ. of Newcastle, New South Wales, Australia.
[2]Senior Lecturer, Dept. of Civil Engg. and Surv., Univ. of Newcastle, New South Wales, Australia.
[3]Postgraduate Student, Dept. of Civil Engg. and Surv., Univ. of Newcastle, New South Wales, Australia.

tests on brick masonry panels. This material model reproduces the inelastic deformations which are typical of brick masonry and incorporates a failure criterion which takes into account the orientation of the jointing planes as well as the state of stress. Details of the program are described and its effectiveness is demonstrated by comparison with experimental results obtained from racking tests on steel frames with brick masonry infills.

2. PROPERTIES OF BRICK MASONRY

To fully define the failure characteristics of brick masonry under biaxial stress, a large number of biaxial tests on half scale brick masonry panels were conducted. From the results, elastic and plastic stress-strain relationships and a criterion of failure were derived. The following is a brief description of the experimental investigation. More detailed descriptions have been reported elsewhere (Page, 1981; Page, 1983).

2.1 Experimental Investigation

Half scale bricks 4.3 in. x 2.0 in. x 1.4 in. (110mm x 50mm x 35mm) were sawn from solid clay paving bricks of dimensions 4.3 in. x 2.0 in. x 9.1 in. (110mm x 50mm x 230mm) and used to construct 14.2 in. (360mm) square brick masonry panels in stretcher bond. The panels were made with a range of angles of the bed joint to the side of the panel (θ). A 1:1:6 mortar (cement:lime:sand, by volume) was used. Consistency in workmanship was achieved by casting the panels horizontally. The panels were moist cured for 28 days before testing.

A biaxial stress state was induced by loading the specimen in orthogonal directions by means of hydraulic jacks. To minimise the effects of platen restraint on the ends of the specimen, flexible brush platens were used on the four loaded edges. Tensile loads were applied by gluing the brush platen to the edges of the panel. Three elongations were measured on each side of the panel using linearly variable differential transducers (LVDTs). All panels were tested under monotonically increasing load with a constant principal stress ratio (σ_1/σ_2). The directions of the stresses, elongations and bed joint angle are as shown in Figure 1.

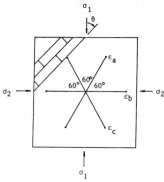

Figure 1 Orientation of Applied Stresses and Measured Strains

2.2 Deformation Characteristics

Masonry is a two phase material in which the mortar is usually more flexible than the brick and generally begins to deform inelastically much earlier than the brick. Most of the non linear deformation in brick masonry, until just before failure, occurs only in the joints.

Joints also influence the failure of brick masonry because they act as planes of weakness. Hence it is appropriate to refer stresses and strains to coordinate axes parallel and perpendicular to the bed joints. The applied stresses (σ_1,σ_2) and the observed strains $(\varepsilon_a,\varepsilon_b,\varepsilon_c)$ were transformed to the stress state (σ_n,σ_p,τ) and the strain state $(\varepsilon_n,\varepsilon_p,\gamma)$ as shown in Figure 2.

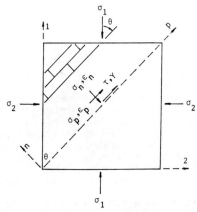

Figure 2 Stress and Strain Transformations

At high stress levels in the compression-compression tests, non linear behaviour was apparent whereas in the tension-compression tests little such behaviour was evident.

2.2.1 Elastic properties

From the initial linear portions of the stress-strain curves, the Young's moduli, Poisson's ratios and shear modulus were determined assuming orthotropic behaviour. Since only three independent strains can be measured for each test, the elastic constants were determined from a regression analysis of the combined results for 112 panels. A summary of the elastic constants is presented in Table 1. Details of the derivation of the constants have been previously reported (Dhanasekar et al, 1982).

For the particular brick masonry used in the tests the results suggested that the elastic behaviour was, on average isotropic. Re-analysis, assuming isotropic behaviour, gave mean values of Young's modulus and Poisson's ratio of 842 ksi (5800 MPa) and 0.23. The respective coefficients of variation were 0.20 and 0.58.

Table 1 Elastic Constants of Brick Masonry

Elastic Property	Mean Value	Standard Error
Young's modulus normal to bed joint, E_n	813 ksi (5600 MPa)	23 ksi (160 MPa)
Young's modulus parallel to bed joint, E_p	827 ksi (5700 MPa)	22 ksi (150 MPa)
Poisson's ratio normal to bed joint, ν_n	0.19	0.021
Poisson's ratio parallel to bed joint, ν_p	0.19	0.020
Shear modulus, G	341 ksi (2350 MPa)	17 ksi (120 MPa)

2.2.2 Non linear stress-strain relations

The biaxial tension-compression tests exhibited little non linearity and were excluded from consideration in determining the parameters in the non linear stress-strain relations. Initially isotropic plastic behaviour was assumed and a yield criterion relating the second invariants of the plastic strains and the deviatoric stresses was derived. This isotropic model was found to consistently overestimate the normal strains (ε_n and ε_p) and underestimate the shear strains (γ). A suitable non-isotropic plastic stress strain relation was then sought.

The following assumptions were found to represent reasonably the average non isotropic plastic behaviour of the panels:

i) plastic strains are related to stress by a power law and are evident at all stress levels,
ii) a plastic strain component related to the jointing directions is, on average, dependent only on the corresponding stress,

and lead to the following relations:

$$\varepsilon_n^p = 10^{-3} (\sigma_n/B_n)^{3.3} \qquad (1)$$
$$\varepsilon_p^p = 10^{-3} (\sigma_p/B_p)^{3.3} \qquad (2)$$
$$\gamma^p = 10^{-3} (\tau/B_s)^{4.0} \qquad (3)$$

in which, the superscript p denotes plastic, and the subscripts n and p refer to the normal and parallel directions. The constants B_n, B_p, B_s have been taken to be 1060 psi (7.3 MPa), 1160 psi (8.0 MPa) and 290 psi (2.0 MPa) respectively, and indicate the stress levels at which the plastic strains are becoming significant.

The variability in the data does not warrant more complex relations. A detailed description of the derivation of these relations has been published recently (Dhanasekar et al, 1985).

2.3 Failure Surface

The main objective of the panel tests was to determine a failure surface for brick masonry subjected to biaxial stresses. The presence of mortar joints in masonry influences its strength and it is necessary to consider the inclination of the bed joint to the principal stresses as one of the variables in defining the failure surface.

2.3.1 Failure surface in σ_1, σ_2, θ space

Each biaxial test for a given value of σ_1, σ_2 and θ represents a point on a general failure surface. A total of 180 tests was performed for various principal stress ratios and bed joint orientations, and a failure surface derived. Detailed discussions on the failure surface may be found elsewhere (Page, 1981; Page, 1983).

The derived failure surface in σ_1, σ_2, θ space is irregular in shape and is difficult to represent algebraically. In addition, local concavities in the biaxial tension-compression quadrant of the failure surface could pose problems in numerical calculations. Thus an alternative failure surface in terms of a stress system parallel and perpendicular to the bed joint was considered.

2.3.2 Failure surface in σ_n, σ_p, τ space

The experimentally derived failure points (σ_1,σ_2,θ) were transformed to (σ_n,σ_p,τ) space. Using the transformed stress points (σ_n,σ_p,τ), shear stress contours were drawn in the σ_n - σ_p plane. The contours suggested that the failure surface could be idealised by three elliptic cones in σ_n,σ_p,τ space. The equations of the cones were derived and plotted in a perspective view in Figure 3.

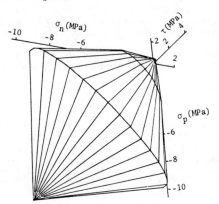

Figure 3 Failure Surface in σ_n, σ_p, τ Space

The equations of the elliptic cones which are symmetric with respect to the σ_n, σ_p plane, in general may be written as

$$C_1\sigma_n^2 + C_2\sigma_p^2 + C_3\tau^2 + C_4\sigma_n\sigma_p + C_5\sigma_n + C_6\sigma_p + 1 = 0 \quad (4)$$

where $(C_i, i = 1,6)$ are constants.

This equation may be expressed in non dimensional form by relating the stresses to suitable reference levels. If the reference levels are the same as those used in the non linear stress-strain relations and non dimensional stress ratios $s_n = \sigma_n/B_n$, $s_p = \sigma_p/B_p$, $t = \tau/B_s$ are introduced, the equations of the cones become

$$c_1 s_n^2 + c_2 s_p^2 + c_3 t^2 + c_4 s_n s_p + c_5 s_n + c_6 s_p + 1 = 0 \quad (4a)$$

The constants c_i for each of the three cones are given in Table 2 and are such that the left hand side of equation (4a) is negative if a stress point lies outside the cone.

Table 2 Constants of the Elliptic Cones

Cone	c_1	c_2	c_3	c_4	c_5	c_6
1	0.0032	-0.0410	-0.0972	0.701	0.869	0.766
2	-14.0	-20.9	-13.0	159.	-13.6	-8.88
3	-1.57	-2.18	-24.6	348.	-19.4	-16.6

Notes: The reference levels are B_n = 1060 psi (7.3 MPa), B_p = 1160 psi (8.0 MPa) and B_s = 290 psi (2.0 MPa)

A detailed description of the derivation of the above surface has been published previously (Dhanasekar et al, 1985a).

It is significant to note that this surface and that recently proposed by Ganz and Thürlimann (1983) for hollow masonry have a number of features in common.

2.3.3 Modes of failure

The separate cones in the idealised failure surface do not correspond to different modes of failure. In view of the need to distinguish in particular between tensile and sliding failures in the joints in the finite element model, the experimental results were re-examined. It was found that whenever there was a tensile stress on a joint, the failure occurred by separation of the joint and no further shear or tensile load could be supported by the joint. There were insufficient results to clearly define the boundary between sliding and crushing failures. It has been assumed that sliding occurs if the normal compressive stress is less than 290 psi (2.0 MPa). The experimental results are consistent with this assumption. During a sliding failure, considerable deformation occurred without appreciable changes in the shear force on the joint. In a crushing failure, all three stress components decreased rapidly.

3. FINITE ELEMENT MODEL

3.1 General

One of the advantages of the proposed material model is that average properties which include the influence of both brick and joint have been derived. This means that a relative coarse finite element mesh can be used with elements encompassing a number of bricks and joints. This has considerable computational advantages when analysing large wall panels.

In general, masonry structures such as shear walls, infilled frames and walls supported on beams, are subjected to in-plane loads and are in a state of plane stress. Rectangular elements are appropriate to such structural elements. A four noded element is somewhat overconstrained for use in regions with moderate stress gradients. An eight noded parabolic element has therefore been adopted.

The eight noded element used in the model is isoparametric with the integrations carried out using four point Gauss quadrature (Owen and Hinton, 1980).

3.2 Program Structure

An elasto-plastic, plane stress finite element program developed by Owen and Hinton (1980) was modified to incorporate the derived material model. In the program the loads are imposed incrementally. For the first (small) increment, elastic behaviour is assumed. Principal stresses are calculated at the Gauss points in each element and if one of these stresses is tensile, the local behaviour is assumed to be elastic-brittle; otherwise elasto-plastic behaviour is assumed. This subdivision of the masonry infill into regions of different behaviour leads to appreciable savings in computation time if the elastic-brittle regions are large. The assumption that the plastic strains are negligible for stress states with some principal tension will be most in error when the stress state is almost uniaxial. Because failure occurred in uniaxial tests before appreciable plastic deformation had taken place, the assumption is not likely to be a critical one.

For each successive load increment two sets of iterations are carried out, one to allow for the material non-linearity and the other to allow cracking to progress. If both principal stresses are compressive and elasto-plastic behaviour is assumed, the stress-strain transformation matrix $[D_{ep}]$ which relates increments in σ_n, σ_p and τ to increments in ε_n, ε_p and γ is given by

$$[D_{ep}]^{-1} = \begin{bmatrix} \frac{1}{E} + \frac{1}{H'_n} & -\frac{\nu}{E} & 0 \\ -\frac{\nu}{E} & \frac{1}{E} + \frac{1}{H'_p} & 0 \\ 0 & 0 & \frac{1}{G} + \frac{1}{H'_s} \end{bmatrix} \quad (5)$$

where H_n', H_p', H_s' are the instantaneous slopes of the stress/plastic-strain relations (1), (2) and (3). Element stiffness matrices are evaluated using $[D_{ep}]$ or its elastic equivalent $[D_e]$. The iterative loop which accounts for the material non linearity ceases when the ratio of the root mean square residual forces to the root mean square external forces is less than a selected tolerance.

The stresses at each Gauss point in each element are then checked for violation of the failure criteria given by equations (4). If failure is indicated, the stress strain relations [D] are modified according to the mode of failure as defined in 2.3.3. The revised relations are shown in Table 3 together with the changes made to the stress state at the Gauss point. The use of reduction coefficients α and β follows the suggestion of Valliappan and Doolan (1972) and provides a means of controlling numerical instability during the development of failure zones. Values adopted for α ranged from 0.002 to 0.1 and for β from 0.02 to 0.05. (The higher values were found necessary when modelling the large deflection stages of infilled frames subjected to racking loads.) The solution process is repeated until no further cracks form and the unbalanced residual forces are within tolerance.

The next load increment is then applied and the procedure repeated. Final failure is indicated by large residual forces, and a lack of convergence in the calculation of deformations.

Table 3 Modification of Stiffness Matrix for Different Modes of Failure

No.	Mode of Failure	Modified [D] Matrix	Stress Components set to zero
1.	Tension failure normal to bed joint	$\begin{bmatrix} \alpha.E & 0 & 0 \\ 0 & E & 0 \\ 0 & 0 & \alpha.G \end{bmatrix}$	Normal stress (σ_n) and shear stress (τ)
2.	Tension failure parallel to bed joint	$\begin{bmatrix} E & 0 & 0 \\ 0 & \alpha.E & 0 \\ 0 & 0 & \alpha.G \end{bmatrix}$	Parallel stress (σ_p) and shear stress (τ)
3.	Shear failure	$\dfrac{E}{1-\nu^2} \cdot \begin{bmatrix} 1 & \nu & 0 \\ \nu & 1 & 0 \\ 0 & 0 & \beta(\frac{1-\nu}{2}) \end{bmatrix}$	None
4.	Biaxial tension or biaxial compression (crushing failure)	$\begin{bmatrix} \alpha.E & 0 & 0 \\ 0 & \alpha.E & 0 \\ 0 & 0 & \alpha.G \end{bmatrix}$	All stress components

4. INFILLED FRAME ANALYSIS

To test the adequacy of the above material model, the finite element program was extended to enable the behaviour of infilled frames up to the stage of failure of the infill to be simulated. The results of five racking tests on infilled frames were thus able to be compared with corresponding finite element results. The program can model the separation of the frame from the infill and the post cracking behaviour of the infilled frames. At present the program does not take account of inelastic behaviour of the surrounding frame nor the existence of clearance between the infill and the frame, although it could be modified to do so.

4.1 Joint and Frame Elements

Two additional finite element types were used in the infilled frame program: a joint element to model the mortar joint between the infill and the frame; and a beam element for the frame. The joint element is a six noded isoparametric one dimensional element and is an extension of the four noded joint element used by Page (1978). Because similar mortars were being modelled, the same values of shear modulus 87 ksi (600 MPa) and Young's modulus 181 ksi (1250 MPa) and the same failure criteria were used. The failure criterion for the element under shear and compression was approximated by a bilinear relation between shear stress and compressive stress. The tensile bond strength between the mortar and the steel frame was low and therefore assumed to be zero.

The frame has been modelled using conventional two noded elastic beam elements with three degrees of freedom at each node. To account for the eccentricity between the centre line of the frame and the outer edge of the mortar joint, the stiffness matrix of the beam element was transformed to suit displacements at the beam-joint interface.

5. VERIFICATION OF THE FINITE ELEMENT MODEL

5.1 Infilled Frame Tests

Racking tests on five infilled frames were used to verify the finite element model. The racking tests were chosen because simple statically determinate boundary conditions could be maintained, and failure of the frames could take place by diagonal cracking or corner crushing of the infill. Thus not only failure loads but failure modes could be verified. Varying stress states and failure modes within the infill were achieved by varying:
(a)   the relative stiffness of the frame and masonry;
(b)   the height/width ratio of the frame;
(c)   the curing history of the masonry and thus its relative bond and compressive strengths.

5.1.1 The test frames

The details of the steel frames are summarised in Table 4. For frames 1, 2 and 5, cold rolled channel sections welded back to back were used; for frame 3, a stiffer tapered flange I section; and for frame 4, an even heavier parallel flange section was selected. The cold rolled sections were reinforced with 8mm web stiffners at the loaded and reaction corners.

Table 4 Details of Infilled Frames

Frame	Length in (mm)	Height in (mm)	I in⁴(10⁶mm⁴)	A in²(mm²)	λh
1,5	61 (1555)	42 (1060)	0.51 (0.212)	0.89 (576)	6.33
2	43 (1095)	42 (1060)	0.51 (0.212)	0.89 (576)	6.44
3	46 (1172)	45 (1137)	10.9 (4.54)	2.65 (1710)	3.24
4	47 (1195)	46 (1160)	30.3 (12.6)	4.62 (2980)	2.37

Also given in the table is the parameter λh first introduced by Stafford Smith (1962) which is a measure of the ratio of the infill stiffness to the frame stiffness. It is defined by:

$$\lambda h = h \left(\frac{E_b \, t \sin 2\theta}{4 \, E_s \, I \, h} \right)^{\frac{1}{4}} \qquad (6)$$

in which E_b, t and h are Young's modulus, thickness and height of the brick masonry infill respectively; E_s and I are Young's modulus and second moment of area of the steel frame section and θ is the angle of the infill diagonal to the horizontal.

The masonry infill was constructed from the same bricks and mortar and using the same casting procedure as in the panels for the biaxial tests. All bed and header joints were 0.2 in. (5mm) with an 0.3 in. (8mm) joint between the infill and the frame. The frames were tested damp at 28 days with the exception of frame 5 which was tested dry at an age of 105 days.

5.1.2 Instrumentation

Electrical resistance strain gauges were attached to the steel frames at sufficient points to allow bending moments and axial forces to be determined at a number of sections. Targets for demountable (Demec) extensometers were glued to the brick masonry infill in 45° rosette arrays at a number of locations. The extensometer gauge length of 7.87 in. (200mm) was sufficient to extend across at least one mortar joint horizontally and several joints vertically. Deflection measurements were taken with mechanical dial gauges around the outside of the frame.

5.1.3 Testing

All five infilled frames were tested under monotonically increasing racking load until the brick masonry infill failed. The load was applied horizontally near the top of one of the frame verticals and reacted at a horizontal support near the bottom of the other vertical. In-plane rotation of the frame was prevented by roller supports at the top of the first vertical and the bottom of the second. The load was increased in 1.12 kip (5 kN) increments until a continuous crack across the panel was readily visible. For frame 4, the loading was continued well past this point to study the behaviour of the frame in the post-cracking range.

In frames 1 to 4, the infill failed down the loaded diagonal in a series of steps along the mortar joints, from the loading to the reaction point. In these cases there was no apparent distress in the bricks. In frame 5, the infill failed by localised crushing near the loaded corner.

5.2 Comparison of Results

All five infilled frame tests were modelled using the finite element program. In simulating each test, 1.12 kip (5 kN) load increments were used. For frames 1 to 4, the material model described in Section 2 was used. For frame 5 the failure surface was modified to account for the different curing history of the infill. Tests on wet and dry specimens (stack bonded prisms for compression; couplets for tensile bond; and triplets for shear bond) revealed that the shear and tensile bond strengths of the brick masonry were approximately 75% greater in the dry specimens than in wet specimens. In contrast, the compressive strength remained unchanged. The relevant sections of the failure surface were modified accordingly.

5.2.1 Load deflection behaviour

A comparison of the observed and predicted load deflection behaviour for each of the five frames is shown in Figure 4.

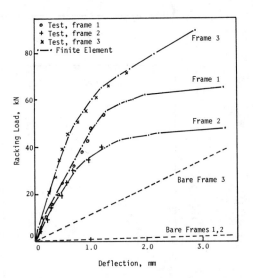

Figure 4a   Load-Deflection Curves for Frames 1, 2, 3
(1 kN = 225 lbf, 1 mm = 0.0394 in.)

12 STRUCTURAL MASONRY ANALYSIS

Figure 4b
Load Deflection
Curve for Frame 4
(1 kN = 225 lbf)
(1 mm = 0.0394 in.)

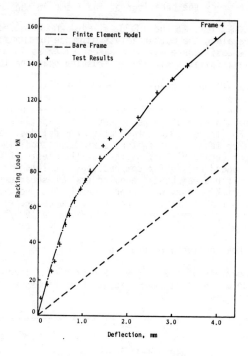

Figure 4c
Load Deflection
Curve for Frame 5
(1 kN = 225 lbf)
(1 mm = 0.0394 in.)

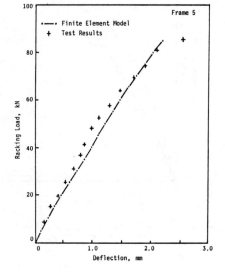

It can be seen from Figures 4(a)-(c) that the finite element model satisfactorily reproduces the load-deflection characteristics of the infilled frames even into the post-cracking range (Figure 4(b)). The load-deflection characteristics are markedly non linear with the slope of the curve, after cracking of the infill, being similar to that of the bare frame. It is also apparent from the curves that despite the fact that the shear failure has progressed along the whole panel, appreciable frictional shear stresses are still transmitted across the failure plane. Comparison of the curves for the three frames with the same height/width ratio (frames 2, 3 and 4) shows the increasing contribution of the infill with increasing frame stiffness. This may be attributed to the increase with frame stiffness of the length of contact between the frame and the infill after separation has taken place. The width of the diagonal compression strut is accordingly increased, together with an increase in its shear bond strength.

5.2.2 Failure of the infill

For frames 1 to 4 the infill failed in a stepped manner from the loading to the reaction point. Only the final cracking pattern could be observed with the naked eye. When the failure region, as predicted by the finite element model, was compared to the observed pattern, good agreement was obtained. For frame 5, failure was confined to a local area near the loaded corner of the infill (under the action of biaxial compressive stresses). The location and extent of the failure zone was again in good agreement. The predicted and observed ultimate loads of the infill are shown in Table 5. Agreement is again satisfactory.

Table 5 Comparison of Failure Loads of the Infill

Frame No.	Observed Load kip (kN)		Finite Element Model kip (kN)		Failure Mode
1	12.5	(55.5)	12.8	(57)	A
2	10.1	(45.0)	9.7	(43)	A
3	14.6	(65.0)	13.7	(61)	A
4	16.9	(75.0)	15.7	(70)	A
5	19.2	(85.5)	19.1	(85)	B

Note: A: Diagonal Splitting   B: Corner Crushing

5.2.3 Other comparisons

Load effects within the frame are more sensitive to modelling assumptions and hence serve as indicators to the adequacy of the model. The bending moment in the frame is affected by the separation of the frame and infill. The variation of bending moment with applied load at two sections in a vertical leg of frame 1, is shown in Figure 5. The effects which influence the bending moments include: the separation of the frame and infill; the extent of cracking and sliding in the joints; and the non linear deformation of the infill. Given the variability of each of these effects, the agreement between the results is satisfactory.

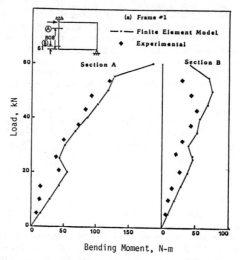

Figure 5 Variation of Bending Moment with Load
(1 kN = 225 lbf, 1 N.m = 0.737 lbf.ft)

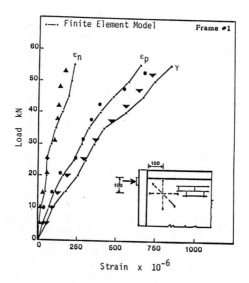

Figure 6 Variation of Strains near the Loaded Corner (1 kN = 225 lbf)

The strains averaged over gauge lengths about a point in the brick masonry infill depend on the length of contact between the frame and the infill, and this in turn is dependent upon the relative stiffness of the frame and the infill. Predicted and measured load/strain curves for frame 1 are shown in Figure 6. The agreement is again satisfactory.

Similar agreement was obtained in other frames. Although experimental values of the bending moments could not be obtained near the loaded corners, in other parts of the frames the variation of bending moment along the frame members at selected load levels were generally in good agreement (Dhanasekar et al, 1985b).

6. THE EFFECT OF VARIATION OF MATERIAL MODEL PARAMETERS

As a result of the good agreement between predicted and observed performance, the finite element model has been used to carry out a study of the importance of a number of the material model parameters in the behaviour of infilled frames subjected to racking loads. Because frames 2 and 5 failed in two distinct modes, namely diagonal cracking and corner crushing, they were used as the basic frames in a study of the effect of varying the brick masonry properties. From the analyses it was found that the modulus of elasticity and the relative bond and compressive strengths were the more important parameters. The behaviour of infilled frames was insensitive to Poisson's ratio and the constants of the inelastic stress-strain relations. In fact for the frames analysed, elastic-brittle material behaviour was found to adequately model the overall behaviour.

The elastic modulus of the masonry (E_b) influences the infilled frame behaviour in several ways. It leads directly to a reduction in racking deflection at a given load, but it also influences the relative stiffness parameter λh which in turn influences the effective width of the "diagonal compression strut" formed in the infill. Increasing E_b without increasing the material strengths leads to a decrease in the failure load of the infill particularly in those cases which are governed by a corner crushing failure.

An increase in compressive strength only influences the failure of the infill when the final failure is due to biaxial compression (corner crushing). It is however possible for the failure mode to change from corner crushing to diagonal cracking for larger increases in compressive strength.

Frame 2 was used to study the influence of varying the shear and tensile bond strengths, both independently and together, with the other material parameters held unchanged. The load at which the infill failed and its failure mode are given in Table 6.

In Table 6 it can be seen that the failure load of the infill is significantly affected in all cases, and that in one case, there is even a change in the failure mode from diagonal cracking to corner crushing. The load /deflection behaviour of the panels was also affected significantly in that the level at which the panel stiffness begins to decrease varies markedly with bond strength.

Table 6 Influence of Bond Strength on the Failure of Brick Masonry Infill

Analysis No.	Tensile Bond Strength psi (kPa)	Shear Bond Strength psi (kPa)	Ultimate Load kip (kN)	Mode of Failure
1	58 (400)	44 (300)	9.7 (43)	A
2	0 0	44 (300)	2.2 (10)	A
3	116 (800)	44 (300)	14.2 (63)	A
4	58 (400)	22 (150)	3.6 (16)	A
5	58 (400)	87 (600)	16.9 (75)	A
6	116 (800)	87 (600)	16.9 (75)	B
7	29 (200)	22 (150)	2.7 (12)	A

Notes: Original material model was used in **Analysis 1**
A: Diagonal Cracking   B: Corner Crushing

The relative importance of compressive strength and the two bond strengths depends upon the failure mode which is in part a function of the ratio of these strengths. The results for test frame 5 brought to light the extent to which the ratio of the bond strengths to compressive strength is influenced by the age and moisture content of the brick masonry. For infilled frames with relatively flexible framing members (i.e. larger values of λh) which are more likely to fail by corner crushing, the mode of failure will be sensitive to the bond strength/compressive strength ratios.

7. CONCLUSIONS

A finite element method of modelling brick masonry subjected to in-plane loading has been developed. The method makes use of a macroscopic continuum model of the stress-strain and failure characteristics of brick masonry which takes into account the influence of the mortar joints acting as planes of weakness.

The material model has been developed from the results of an extensive set of biaxial tests on brick masonry panels in which the angle of the bed joint to the edges of the panel were varied. The properties are most conveniently expressed in terms of stresses and strains normal and parallel to the bed joint planes and in terms of bed joint shear stresses and strains.

In the elastic range, the panels tested were found to be on average almost isotropic and isotropic elastic behaviour is assumed. In the plastic range up to loads just below the failure load, most of the plastic deformation appears to be associated with the joints and is distinctly orthotropic. There is little interdependence of the normal, parallel and shear strains with the non-corresponding stresses. Simple power law relations between the above plastic strains and the corresponding stresses were able to be assumed.

A failure criterion for the brick masonry panels was also derived in terms of the normal, parallel and joint shear stresses. The failure surface has been idealised as a surface consisting of three elliptic cones in the σ_n, σ_p and τ space.

In the finite element model, there is a need to distinguish between tensile failures, bed joint sliding failures and crushing failures. In the model, if the failure surface is reached and there is tension on a joint the failure is assumed to be a tensile failure; if the failure surface is reached and the stress normal to the bed joint is compressive and less than 290 psi (2 MPa) the failure is a shear failure. Otherwise the failure is assumed to occur by splitting in the plane of the panel.

The finite element model has been used to predict the behaviour of five rectangular infilled steel frames for which test results were available. The load deflection behaviour, the modes of failure, failure loads and "local" stress-strain behaviour are all in good agreement.

The finite element model has also been used to determine which of the material model parameters are more critical in predicting the behaviour of infilled frames. They include, the elastic modulus of brick masonry, the compressive strength, the shear and tensile bond strengths of the masonry. For the infill frames studied, the influence of the parameters in the stress/plastic-strain relations was small and the assumption of elastic-brittle behaviour throughout the infill is justifiable.

It should be noted that the finite element model makes use of material properties derived from tests on solid brick masonry. It may also be used in the analysis of hollow core masonry if appropriate material properties are used.

8. ACKNOWLEDGEMENTS

The assistance of the laboratory staff of the Department of Civil Engineering and Surveying, the University of Newcastle, in carrying out the experimental investigation is gratefully acknowledged. Part of the research was financed by the Australian Research Grants Scheme. Bricks were supplied by PGH Ceramics - Bricks, N.S.W.

9. REFERENCES

1) Dhanasekar, M., Page, A.W., and Kleeman, P.W., "The Elastic Properties of Brick Masonry", *International Journal of Masonry Construction*, Vol. 2, No. 4, 1982, pp. 155-160.

2) Dhanasekar, M., Kleeman, P.W., and Page, A.W., "Non Linear Biaxial Stress-Strain Relations for Brick Masonry", *Journal of Structural Division*, ASCE, Vol. 111, No. ST5, Proc. May 1985.

3) Dhanasekar, M., Page, A.W., and Kleeman, P.W., "The Failure of Brick Masonry under Biaxial Stresses", *Proceedings of Institution of Civil Engineers*, Part II, Vol. 79, Paper No. 8871, March (1985a).

4) Dhanasekar, M., Page, A.W., and Kleeman, P.W., "The Behaviour of Brick Masonry under Biaxial Stress with Particular Reference to

Infill Frames", *Seventh International Brick Masonry Conference*, Vol. 2, Melbourne 1985(b), pp. 815-824.

5) Ganju, T.N., "Non Linear Finite Element Analysis of Clay Brick Masonry", *Proceedings of Sixth Australasian Conference on Mechanics of Structures and Materials*, August 1977, Christchurch, pp. 59-65.

6) Ganz, H.R., and Thürlimann, B., "Strength of Brick Walls under Normal Force and Shear", *8th International Symposium on Load Bearing Brickwork*, B.C.R.A., London, November 1983.

7) Owen, D.R.J., and Hinton, E., *"Finite Elements in Plasticity - Theory and Practice"*, Pineridge Press Ltd., Swansea, U.K.

8) Page, A.W., "A Finite Element Model for Masonry", *Journal of Structural Division*, ASCE, Vol. 104, No. ST8, August 1978, pp. 1267-1285.

9) Page, A.W., "The Biaxial Compressive Strength of Brick Masonry", *Proceedings of Institution of Civil Engineers*, Part II, Vol. 71, September 1981, pp. 893-906.

10) Page, A.W., "The Strength of Brick Masonry under Biaxial Tension-Compression", *International Journal of Masonry Construction*, Vol. 3, No. 1, 1983, pp. 26-31.

11) Samarasinghe, W., Page, A.W., and Hendry, A.W., "A Finite Element Model for the Inplane Behaviour of Brickwork", *Proceedings of Institution of Civil Engineers*, Part II, September 1981, pp. 171-178.

12) Stafford-Smith, B., "Lateral Stiffness of Infilled Frames", *Journal of Structural Division*, ASCE, Vol. 88, No. ST6, Proc. 1962, pp. 183-189.

13) Valliappan, S., and Doolan, T.F., "Non Linear Stress Analysis of Reinforced Concrete", *Journal of Structural Division*, ASCE, Vol. 98, No. ST4, Proc. 1972, pp. 885-898.

NONLINEAR MACROSCOPIC FINITE ELEMENT MODEL FOR MASONRY WALLS

Ahmed S. Essawy,[1] Robert G. Drysdale,[2] and Farooque A. Mirza[3]

A macroscopic finite element model which includes both transverse shear effects and nonlinearity due to cracking has been developed for masonry. These features are accommodated using a multi-layered model. This model can handle general in-plane and out-of-plane loading conditions simultaneously. It is capable of tracing the crack progress and cracking pattern in the plane of the wall and modifying the stiffness of the cracked elements only in the cracked layers. These layers can correspond to face shell(s) in a hollow block wall or wythe(s) in the case of a multiwythe wall.

INTRODUCTION

Background: Initially, finite element modelling of large scale structural masonry assemblages employed standard linear elastic concepts. In these models average properties used in the context of assumed isotropic elastic behavior greatly simplified their development. However this form of analysis cannot fully recognize the presence of mortar joints as particular planes of weakness and it ignores the anisotropic and composite nature of masonry. Also this simple method of analysis is only applicable up to first cracking so that any reserve of strength after first cracking is neglected. Moreover, for out-of-plane bending, the transverse shear deformations which have been shown to be significant for hollow block walls are ignored. This subject is discussed in more detail later in this paper.

---

[1] Ph.D. Candidate, Department of Civil Engineering and Engineering Mechanics, McMaster University, Hamilton, Ontario, Canada
[2] Professor, Department of Civil Engineering and Engineering Mechanics, McMaster University, Hamilton, Ontario, Canada
[3] Associate Professor, Department of Civil Engineering and Engineering Mechanics, McMaster University, Hamilton, Ontario, Canada

To include the presence of mortar joints as planes of weakness, a more microscopic approach was proposed (3) using different continuous elements to separately model the mortar joint and masonry unit materials. However, the bond strengths of the mortar-unit interface were used for the shear and tensile strengths of the mortar joint material. Since masonry usually experiences failure by debonding along mortar joints, this procedure avoided modelling the mortar-unit interface.

Recently, another microscopic model has been proposed (2) to model masonry as a discontinuous system where the discontinuities consist of the mortar joints. Besides the separate modelling of mortar and masonry unit materials (as well as grout and steel reinforcement, if present), the physical behaviour of the interfaces between different materials was added by introducing double node pairs (one on each side of each interface). The interconnection between the double nodes was specified to simulate the interface behavior.

A third microscopic model was developed (12,13) from the analogy of the behavior of masonry assemblages and jointed rock (8). In this approach, the masonry units were modelled as linear elastic continuous elements and the mortar joints were modelled as linkage or joint elements.

The microscopic models mentioned above accounted for the composite nature of masonry and considered the inherent nonlinearity due to both mortar joint behavior and progressive joint failure. However, application of these models for the analysis of large structural masonry elements and structures is extremely difficult due to the very large number of elements needed to separately model the component materials and their interfaces. Moreover, the development of these models was limited to in-plane loading cases. Also, the anisotropic nature of the masonry units was not incorporated into these models.

Scope: It is the purpose of this paper to report a recently developed macroscopic finite element model which can be used to efficiently predict the capacity and behavior of full scale masonry walls and large structural elements. In this model, the masonry element is discretized to finite elements without particular regard to the position of the mortar joint planes. This means that some mortar joint will be included within the macro-element. However, the failure criteria implemented in this model consider not only the occurrence of failure but also the failure pattern with respect to the mortar joints.

The proposed model has been developed in a generalized form to handle both in-plane and out-of-plane loading conditions so that it can be used for analysing most masonry assemblages. This feature was achieved by incorporating both in-plane and out-of-plane degrees of freedom in the element formulation.

SIGNIFICANCE OF TRANSVERSE SHEAR DEFORMATIONS IN HOLLOW BLOCK MASONRY

The longitudinal section parallel to the bed joints of a hollow concrete block wall has the appearance of a Vierendeel truss. When these walls are subjected to out-of-plane bending about the vertical axis, some significant transverse shear deformation may be experienced because of the presence of the relatively flexible webs connecting the face shells. In order to judge the significance of the shear deformations in the case of hollow concrete masonry, a longitudinal strip of a masonry wall supported along the two vertical sides was analysed using three different methods. First, it was analysed as a Vierendeel truss. This was considered to represent the actual behavior of this strip. Second, a beam theory analysis, which corresponds to the case of negligible shear deformations was done. Third, a sandwich panel (1) analysis was employed, in which the cross webs were replaced by an equivalent shear lamina. In all of these analyses, representative values were used for the block and mortar properties. Also, uniformly distributed loading was specified and relatively large span to depth ratios were used. Both the uniformly distributed load and the long spans tend to minimize any transverse shear effects. However the Vierendeel truss analysis showed that neglecting the transverse shear resulted in significantly less deflections. For instance, ignoring shear deformations for a 5 m span and standard 190 mm thick hollow concrete blocks resulted in 14% less deflection. Modelling the section using the sandwich panel approach with an equivalent core layer gave slightly smaller deflections. For the above case these were 6% less than predicted using the Vierendeel model. Accordingly, it was decided that transverse shear deformations can be significant in the case of laterally loaded hollow block masonry and that the cross webs may be replaced by an equivalent lamina or core layer with an acceptable accuracy.

FINITE ELEMENT FORMULATION

Transverse shear deformations are known to have significant effects on behavior of sandwich panels (1) and cellular plates (4) and, as shown in the previous section, even on the horizontal bending of hollow block walls. Therefore, in the formulation of the plate type masonry element, the assumptions adopted by Mindlin (11) are used.

These include that deflections of the plate are small, stresses normal to the middle surface are negligible irrespective of the loading, and normals to the middle surface before deformation remain straight but not necessarily normal to the middle surface after deformation. As discussed elsewhere (9,10,14), the last assumption is not completely valid since some warping of the type shown in Figure 1 does occur as part of the shear deformation. However, a correction can be made to partially allow for non-uniform shear distribution. Therefore, in Figure 1, θ_x and θ_y can be considered as the total section rotations about y and x axes, respectively, where

$$\begin{Bmatrix} \theta_x \\ \theta_y \end{Bmatrix} = \begin{Bmatrix} \frac{\partial w}{\partial x} + \phi_x \\ \frac{\partial w}{\partial y} + \phi_y \end{Bmatrix} \qquad (1)$$

with w as the transverse deflection in the z direction, and ϕ_x and ϕ_y are the uniform shear deformations in the x - z and y - z planes, respectively.

To accomodate both in-plane and out-of-plane loading and displacements, it was decided that the masonry finite element should include all in-plane and out-of-plane (including transverse shear) degrees of freedom. These were included by combining the 8 degree of freedom (DOF) rectangular plane stress element, the 12 DOF nonconforming rectangular plate bending element, and the 8 DOF rectangular transverse shear element. This resulted in the 28 DOF element shown in Figure 2 in which the different displacement fields of the middle surface are given. For the displacement field through the thickness of the plate, the following relations are applicable:

$$\{u\} = \begin{Bmatrix} u(x,y,z) \\ v(x,y,z) \\ w(x,y) \end{Bmatrix} = \begin{Bmatrix} u^o(x,y) - z\theta_x(x,y) \\ v^o(x,y) - z\theta_y(x,y) \\ w(x,y) \end{Bmatrix} \qquad (2)$$

where u, v, and w are the displacements in the coordinate directions, x, y, and z, respectively, and u^o and v^o are the corresponding displacements of the middle surface.

Using the displacments in Equation 2 and following the small displacement assumption, the strains can be written as

MASONRY WALLS MODEL 23

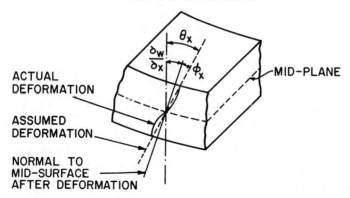

Figure 1. Deformation of the Cross Section of a Homogeneous Plate (9).

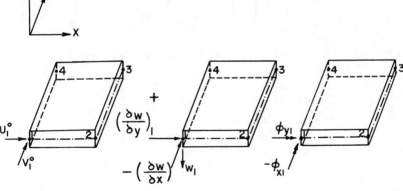

EXTENSION(8 DOF) + FLEXURE (12 DOF) + SHEAR (8 DOF)

$u^o = [1,x,y,xy][a_i]$
$(i = 1,4)$

$v^o = [1,x,y,xy][a_j]$
$(j = 5,8)$

$w = [1,x,y,x^2,xy,y^2,x^3,$
$\quad x^2y, xy^2, y^3, x^3y, xy^3][a_k]$
$(k = 9,20)$

$\theta_x = \frac{\partial w}{\partial x} + \phi_x$

$\theta_y = \frac{\partial w}{\partial y} + \phi_y$

$\phi_x = [1,x,y,xy][a_l]$
$(l = 21,24)$

$\phi_y = [1,x,y,xy][a_m]$
$(m = 25,28)$

Figure 2. Displacement Fields for Rectangular Plate Elements (28 DOF).

$$\{\varepsilon\} = \left\{\frac{\{\varepsilon_{ef}\}}{\{\varepsilon_s\}}\right\} = \left\{\begin{array}{c} \varepsilon_x^o - z(\frac{\partial^2 w}{\partial x^2} + \frac{\partial \phi_x}{\partial x}) \\ \varepsilon_y^o - z(\frac{\partial^2 w}{\partial y^2} + \frac{\partial \phi_y}{\partial y}) \\ \gamma_{xy}^o - z(2\frac{\partial^2 w}{\partial x \partial y} + \frac{\partial \phi_x}{\partial y} + \frac{\partial \phi_y}{\partial x}) \\ \hline -\phi_x \\ -\phi_y \end{array}\right\} \quad (3)$$

where ε_x^o, ε_y^o, and γ_{xy}^o are the middle surface strains and $\{\varepsilon_{ef}\}$ and $\{\varepsilon_s\}$ are the extension/flexure and the shear components of strain, respectively.

The curvatures $\{\chi\}$ including the transverse shear are then given by

$$\{\chi\} = \left\{\begin{array}{c} \chi_x \\ \chi_y \\ \chi_{xy} \end{array}\right\} = \left\{\begin{array}{c} -(\frac{\partial^2 w}{\partial x^2} + \frac{\partial \phi_x}{\partial x}) \\ -(\frac{\partial^2 w}{\partial y^2} + \frac{\partial \phi_y}{\partial y}) \\ -(2\frac{\partial^2 w}{\partial x \partial y} + \frac{\partial \phi_x}{\partial y} + \frac{\partial \phi_y}{\partial x}) \end{array}\right\} \quad (4)$$

and lead to the following modified strain expression

$$\{\varepsilon\} = \{\frac{\varepsilon_{ef}}{\varepsilon_s}\} = \left\{\begin{array}{c} \varepsilon_x^o + z\,\chi_x \\ \varepsilon_y^o + z\,\chi_y \\ \gamma_{xy}^o + z\,\chi_{xy} \\ \hline -\phi_x \\ -\phi_y \end{array}\right\} \quad (5)$$

and $\quad \{\varepsilon_{ef}\} = \{\varepsilon^o\} + \{\varepsilon_f\} = \begin{Bmatrix} \varepsilon_x^o \\ \varepsilon_y^o \\ \gamma_{xy}^o \end{Bmatrix} + z \begin{Bmatrix} \chi_x \\ \chi_y \\ \chi_{xy} \end{Bmatrix}$ (6)

Now, using the displacement fields as shown in Figure 2 for the rectangular plane stress, rectangular, transverse shear elements mentioned before, i.e.

$$u^o(\xi,\eta) = \sum_{i=1}^{4} N_i^o(\xi,\eta) u_i^o$$

$$v^o(\xi,\eta) = \sum_{i=1}^{4} N_i(\xi,\eta) v_i^o$$

$$w(\xi,\eta) = \sum_{i=1}^{12} N_i(\xi,\eta) w_i \quad (7)$$

$$\phi_x(\xi,\eta) = \sum_{i=1}^{4} N_i^s(\xi,\eta) \phi_{x(i)}$$

$$\phi_y(\xi,\eta) = \sum_{i=5}^{8} N_i^s(\xi,\eta) \phi_{y(i-4)}$$

in terms of the non-dimensional ξ, η axes shown in Figure 3, the respective strains can be easily derived via the derivatives of the shape functions N_i^o, N_i and N_i^s as required in Equation 3 or 5. The resulting strain matrix $[\bar{B}]$ then relates the nodal degrees of freedom u_i^o, v_i^o, w_i, ϕ_{xi} and ϕ_{yi} at node i to the strain vector $\{\varepsilon\}^T = \langle \{\varepsilon_{ef}\}^T \{\varepsilon_s\}^T \rangle$.

The derivation of the element stiffness matrix then follows the standard procedure (20) but with the use of the generalized constitutive relations in which the stresses represent the internal forces and the strains are caused by the deformations at the middle surface. This formulation has been used previously for plate bending elements (9,10) and for laminated anisotropic plates (14,15). Accordingly, the element stiffness matrix is given by

$$[K_e] = \int [\bar{B}]^T [\bar{D}] [\bar{B}] \, dA \quad (8)$$
$\quad\;\; 28\times28 \quad A \;\; 28\times8 \;\; 8\times B \;\; 8\times28$

where $[\bar{B}]$ is the generalized strain matrix and $[\bar{D}]$ is the material property or the rigidity matrix. The element stiffness matrix, $[K_e]$, is calculated using the three-point Gauss quadrature integration scheme along each axis. The formulation was checked by performing an eignvalue analysis of the stiffness matrix $[K_e]$. This resulted in

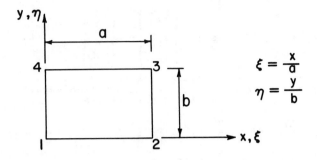

Figure 3. Typical Rectangular Plate Element.

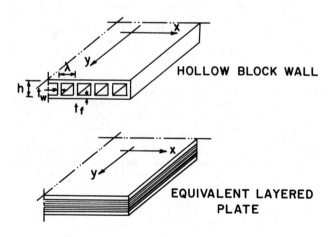

Figure 4. Equivalent Core Layer.

MASONRY WALLS MODEL   27

six zero eignvalues and analysis of the corresponding eigen- vectors indicated that these modes are the rigid body modes both in-plane and out-of-plane. Consequently the validity of the zero strain modes was confirmed.

DEVELOPMENT OF THE RIGIDITY MATRIX

<u>Formulation of the Rigidity Matrix</u>: The rigidity matrix for masonry constructed with solid units can be developed as a single layer using equivalent properties for the masonry material based on experimental results. However, for hollow block or multi-wythe masonry, the rigidity matrix was developed using the concept of a layered plate. In this case, properties equivalent to the face shell and mortar combination were provided for the outer shell (or wythe) and an equivalent core layer was used to replace the cross webs in the case of hollow block construction.

The concept of the layered plate, introduced here, enabled the analysis to be extended beyond the elastic or cracking limit. That is, it was only necessary to modify the stiffness of the particular layer which had reached the elastic limit or cracked condition. This approach permits modelling of observed behavior of laterally loaded hollow block walls (5,6) where cracking was only observed on the tension side of the wall.

The rigidity matrix is derived from the constitutive relations shown below for a layer k.

$$\begin{Bmatrix} \sigma_x \\ \sigma_y \\ \tau_{xy} \\ \hline \tau_{xz} \\ \tau_{yz} \end{Bmatrix}^{(k)} = \begin{bmatrix} (\frac{E_x}{1-\nu_{xy}\nu_{yx}}) & (\frac{\nu_{xy}E_y}{1-\nu_{xy}\nu_{yx}}) & 0 & 0 & 0 \\ (\frac{\nu_{xy}E_y}{1-\nu_{xy}\nu_{yx}}) & \frac{E_y}{(1-\nu_{xy}\nu_{yx})} & 0 & 0 & 0 \\ 0 & 0 & G_{xy} & 0 & 0 \\ \hline 0 & 0 & 0 & G_{xz} & 0 \\ 0 & 0 & 0 & 0 & G_{yz} \end{bmatrix} \begin{Bmatrix} \varepsilon_x \\ \varepsilon_y \\ \gamma_{xy} \\ \hline \gamma_{xz} \\ \gamma_{yz} \end{Bmatrix}^{(k)} \quad (9)$$

Or $\{\sigma\}^k = [D]^k \{\varepsilon\} = \begin{bmatrix} [D_{ef}] & [0] \\ [0] & [D_s] \end{bmatrix} \{\varepsilon\}$

As shown above, the elasticity matrix $[D]^k$ is divisible into the uncoupled extension/flexure and the shear components.

The stress resultants can be obtained by the appropriate integration of the stress components over the layer thickness. Then, summing the stress resultants for all the layers as follows:

$$\{N\} = \begin{Bmatrix} N_x \\ N_y \\ N_{xy} \end{Bmatrix} = \Sigma_{k=1}^{n} \int^k \{\sigma\}^k \, dz$$

$$= \Sigma_{k=1}^{n} \int^k [D_{ef}]^k \{\varepsilon\} \, dz. \qquad (10)$$

Substituting for $\{\varepsilon\}$ from Equation 6 gives

$$\{N\} = [\Sigma_{k=1}^{n} \int^k [D_{ef}]^k dz]\{\varepsilon^o\} + [\Sigma_{k=1}^{n} \int^k [D_{ef}]^k \cdot z\, dz]\{\chi\}$$

or $\quad \{N\} = [\bar{D}_e]\{\varepsilon^o\} + [\bar{D}_c]\{\chi\} \qquad (11)$

where

$$[\bar{D}_e] = \Sigma_{k=1}^{n} \int^k [D_{ef}]^k \, dz \text{ and } [\bar{D}_c] = \Sigma_{k=1}^{n} \int^k [D_{ef}]^k z \, dz.$$

Similar expressions can be derived for $\{M\}$ and $\{Q\}$ and the integrated or generalized constitutive equation may be written as

$$\begin{Bmatrix} \{N\} \\ \{M\} \\ \{Q\} \end{Bmatrix} = \begin{bmatrix} [\bar{D}_e] & [\bar{D}_c] & 0 \\ [\bar{D}_c] & [\bar{D}_f] & 0 \\ 0 & 0 & \alpha[\bar{D}_s] \end{bmatrix} \begin{Bmatrix} \{\varepsilon^o\} \\ \{\chi\} \\ \{\phi\} \end{Bmatrix} \qquad (12)$$

or $\quad \{\bar{\sigma}\} = [\bar{D}]\{\bar{\varepsilon}\}$

where $\quad [\bar{D}_f] = \Sigma_{k=1}^{n} \int^k [D_{ef}]^k z^2 dz \qquad (13)$

$$[\bar{D}_s] = \Sigma_{k=1}^{n} \int^k [D_s]^k \, dz$$

and α is the factor introduced to account for the non-uniform shear deformations or warping of the section. It has been set equal to the previously (10,14) established value of 5/6 for rectangular sections.

It should be noted that the extension/flexure coupling rigidity sub-matrices $[\bar{D}_c]$ will be cancelled for a symmetric layer arrangement. However, it is presented to accomodate asymmetric arrangements.

<u>Calculation of Equivalent Elastic Constants for Different Layers</u>: The face shell layers or the outer wythe layers have been assumed to have identical orthotropic material properties for which the required six elastic constants E_x, E_y, ν_{xy}, G_{xy}, G_{xz} and G_{yz} can be obtained using the available experimental data. However, the cross webs, in the case of hollow block walls, are replaced by an equivalent solid layer as shown in Figure 4. The elastic constants for this layer can be determined by separating

the rigidity expressions derived by Basu and Dawson (4) for Cellular plates into the contributions of the individual layers. Accordingly the elastic constants for the web layer are as follows

$$E_x = 0.0,$$

$$E_y = E_w \frac{t_w}{\lambda},$$

$$\nu_{yx} = \nu_w,$$

$$G_{xy} = 0.0 \text{ for in-plane shear calculations}$$

and
$$G_{xy} = \frac{6 G_f \cdot \frac{t_f}{h}}{(1 - \frac{tf}{h})^3} \left[\frac{1 - \frac{h}{L} \cdot \frac{t_f}{t_w}}{1 + \frac{h}{L} \cdot \frac{t_f}{t_w}} - \frac{1}{3} (\frac{t_f}{h})^2 \right] \quad (14)$$

for out-of-plane twisting calculations.

Also
$$G_{xz} = G_w \frac{1}{1 + \frac{(h-t_f)^2}{2.4(1+\nu_w)t_w^2}}$$

and
$$G_{yz} = G_w \times \frac{t_w}{\lambda}$$

where, E_w, G_w and ν_w are the modulus of elasticity, shear modulus and Poisson's ratio, respectively, for the web material (concrete blocks in this case). In addition, G_f is the shear modulus for the face shell material, L is the transverse span of the panel, and all other notation refers to the transverse section dimensions shown in Figure 4.

It should be noted that the elastic constants mentioned above also satisfy the thermodynamic constraints that lead to the positive definite condition of the rigidity matrix in Equation 9.

MATERIAL NONLINEARITY AND INCREMENTAL ANALYSIS

In this model, the masonry material has been assumed to behave in a linear elastic manner. This is considered accurate for plain masonry where tension cracking is the controlling mode of failure. Therefore the nonlinearity in the analysis is due to the progressive cracking of the assemblage either by debonding at mortar joints or tensile cracking of units or both.

Cracking occurs when the state of stress exceeds the preset conditions for failure according to failure criteria suggested for masonry assemblages in biaxial states of stress (7). Using these criteria, both the occurrence of failure and the failure mode are predicted according to the state of stress and the component materials geometry and properties. After the cracking limit has been exceeded in a certain layer, an incremental stress-strain relationship, similar to the generalized Hooke's law but valid beyond the elastic limit is adopted. It is given by

$$\{d\sigma\} = [D_{cr}]\{d\varepsilon\} \qquad (15)$$

where $\{d\sigma\}^T = \{d\sigma_x \ d\sigma_y \ d\tau_{xy} \ d\tau_{xz} \ d\tau_{yz}\}$ is the stress increment vector, $\{d\varepsilon\}^T = \{d\varepsilon_x \ d\varepsilon_y \ d\gamma_{xy} \ d\gamma_{xz} \ d\gamma_{yz}\}$ is the total strain increment vector, and $[D_{cr}]$ is the modified elasticity matrix for the cracked layer.

Elasticity Matrix for Cracked Layers $[D_{cr}]$: The modification of the elasticity matrix after cracking depends upon the failure mode (or cracking pattern within the cracked layer of the element) and the applied state of stress. The failure modes in the plane of the assemblage were classified under one of three modes shown in Figure 5 which are as follows:

1. Horizontal failure plane, where failure occurs by debonding along a bed joint through the element.

2. Vertical or toothed vertical failure plane, where failure occurs either by debonding along head joints and splitting through masonry units in alternate courses [Mode 2(I)] or by debonding along head and bed joints around the units in a toothed pattern [Mode 2(II)], respectively.

3. Diagonal or diagonal stepped failure plane where failure occurs either by splitting through the units in an almost straight line [Mode 3(I)] or by debonding along bed and head joints around the units in a stepped pattern [Mode 3(II)], respectively.

For cracking Modes 1 and 2, the elasticity matrix given in Equation 8 is modified by reducing the modulus of elasticity in the direction normal to the crack by a reduction factor, α_c, and the shear modulii by a reduction factor, β_c. However, the modulus of elasticity and shear modulus parallel to the failure or cracking plane is left untouched. Accordingly, considering the x and y coordinate to coincide with the parallel and normal directions to the bed joints, respectively, the modified

MASONRY WALLS MODEL

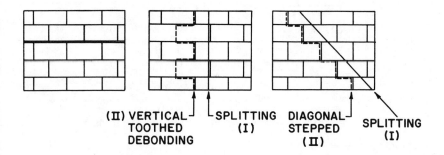

Figure 5. Possible In-Plane Failure Modes.

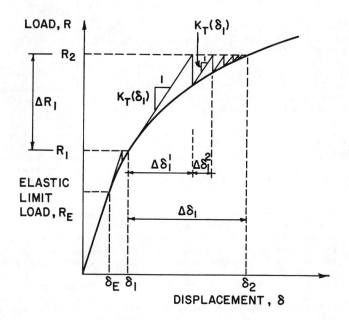

Figure 6. Modified Newton-Raphson Iterative Method.

elasticity matrix for the first failure mode can be written as follows:

$$[D_{cr}]_1 = \frac{1}{1-\nu_{xy}\nu_{yx}} \begin{bmatrix} E_x & \nu_{xy}\alpha_c Ey & 0 & 0 & 0 \\ \nu_{xy}\alpha_c Ey & \alpha_c Ey & 0 & 0 & 0 \\ 0 & 0 & (1-\nu_{xy}\nu_{yx})\beta_c G_{xy} & 0 & 0 \\ 0 & 0 & 0 & (1-\nu_{xy}\nu_{yx})G_{xz} & 0 \\ 0 & 0 & 0 & 0 & (1-\nu_{xy}\nu_{yx})\beta_c G_{y\bar{z}} \end{bmatrix} \quad (16)$$

and for the second failure modes, it is given by:

$$[D_{cr}]_2 = \frac{1}{1-\nu_{xy}\nu_{yx}} \begin{bmatrix} \alpha_c E_x & \nu_{xy}\alpha_c Ey & 0 & 0 & 0 \\ \nu_{xy}\alpha_c Ey & Ey & 0 & 0 & 0 \\ 0 & 0 & (1-\nu_{xy}\nu_{yx})\beta_c G_{xy} & 0 & 0 \\ 0 & 0 & 0 & (1-\nu_{xy}\nu_{yx})\beta_c G_{xz} & 0 \\ 0 & 0 & 0 & 0 & (1-\nu_{xy}\nu_{yx})G_{y\bar{z}} \end{bmatrix} \quad (17)$$

For the third failure mode, factors α_c and β_c are applied for both elasticity modulii and all shear modulii. In the case of reinforced concrete, the values of α_c and β_c vary from 0 to 1.0 depending upon reinforcement dowel action, aggregate interlock, and crack width (13). However for unreinforced masonry, these reduction factors are considered to depend only on the state of stress and the failure mode as follows:

1. For the condition of cracking and tensile stress normal to the crack, α_c is set equal to zero because cracked elements cannot transmit tension normal to the crack. For compression α_c remains equal to unity because compression can be transmitted normal to the crack.

2.(a) For the case of tensile stress normal to a crack resulting from debonding at the interface of the mortar and the units, β_c is set equal to zero. The reason is that no shear can be transmitted across these cracks because the relatively smooth surfaces will not be in contact. However, for tension cracking through the units, it is assumed that the irregular surface of the crack should be able to transfer a certain amount of shear. Accordingly, a value of β_c = 0.4, as previously suggested (3) for concrete blocks, is assumed for cracking through masonry units as in the diagonal splitting mode of failure shown as Mode 3(I) in Figure 5). Where cracking is partially through units and partially along mortar joints as in Mode 2(I) in Figure 5, the average value of $\beta_c = 0.2$ is used.

(b) The compression-shear bond strength before cracking can be represented by the Coulomb-Mohr failure theory (3). After cracking by debonding, shear can only be resisted by friction. Then, the shear reduction factor, β_c, can be determined by

$$\beta_c = \frac{\mu\,\sigma_c}{\mu\,\sigma_c + \sigma_{so}} \qquad (18)$$

where σ_c is the applied compressive stress normal to the crack, σ_{so} is the cohesion or the shear bond strength at zero precompression and μ is the coefficient of friction between mortar and units.

However if cracks run through units a value of $\beta_c = 1.0$ is used and if cracks go through both head joints and masonry units, an average value between 1.0 and the value from Equation 18 is assumed.

<u>Nonlinear Incremental Analysis</u>: Nonlinear finite element analysis is performed in an incremental manner using the modified Newton-Raphson interactive scheme within each load increment. In other words the initial stiffness matrix is held the same for iterations within a load increment. This stiffness matrix is updated after each load increment. At the beginning of each load increment, the tangential stiffness matrix is formulated according to the stress level at the end of the previous load increment. Then, for the present load increment, incremental displacements, strains, and stresses can be determined. The nonlinearity implies that for the calculated increment of strain the computed stress increment, in general, will not be the correct stress increment. The difference between both stress increments or the error is treated as an initial stress which is balanced by revising the correction load increment until the unbalanced forces are within a predetermined tolerance. A graphical representation of this incremental nonlinear analysis is shown in Figure 6.

BOUNDARY CONDITIONS

Since this formulation includes transverse shear deformations, three out-of-plane boundary conditions have to be specified at each edge rather than the two conditions for ordinary thin plate theory. The third required boundary condition is related to the twisting of the edge and it is affected by the edge conditions as follows:

1. If there is an edge stiffener or diaphragm at the edge, then the rotation about an axis normal to that edge has to be restrained to zero (i.e., $\frac{\partial w}{\partial s}$ and ϕ_s are zero).

2. If no edge stiffeners or diaphragms are provided, then the twisting moment about an axis normal to that edge has to be zero. The rotations are not restrained.

Since the finite element formulation is based entirely on assumed displacment functions, only the boundary displacement conditions can be satisfied exactly according to the edge support condition and the edge stiffening conditions. For example for an edge parallel to the y axis, the different possible boundary conditions are as follows:

(i)   For a simply supported edge
   a) with stiffeners $w = 0$, $\theta_y = 0$, $\phi_y = 0$
   b) without stiffeners $w = 0$

(ii) For a clamped edge
   a) with stiffeners $w = 0$, $\theta_x = 0$, $\phi_x = 0$,
                           $\theta_y = 0$, $\phi_y = 0$
   b) without stiffeners $w = 0$, $\theta_x = 0$, $\phi_x = 0$

(iii) For a free edge $\phi_x = 0$

It should be noted that these boundary conditions are related to the out-of-plane degrees of freedom only. However, other boundary conditons for in-plane degrees of freedom could be added if required.

MODEL VERIFICATION USING AVAILABLE SOLUTIONS

Before using the proposed model for predicting the strength of masonry assemblages, a verification study was conducted to check the model predictions against available linear elastic solutions. Included were different cases of thin and thick simply supported plates, isotropic and orthotropic single layer plates, and a multicell bridge deck (as a layered plate having cross webs). Also a convergence study for thin plates was performed for the proposed model. In addition, to verify the nonlinear analysis of the model, a simply supported thin plate problem for which the nonlinear analysis is available in the literature was analysed. These case studies are discussed individually below:

Single layer Isotropic Plates: In order to verify the model for thick plate predictions, simply supported square plates subject to uniformly distributed load and having thickness to side length ratios of 0.01 to 0.25 were analyzed using the proposed model. The results are compared in Figure 7 with two available solutions which include transverse shear deformations for thick plates (14). This figure shows generally good agreement with a maximum difference of 5% for the very thick plate considered.

Single Layer Orthotropic Thin Plates: Simply supported orthotropic thin plates having different side lengths and ratios of orthotropy were analysed and compared to the available solutions in the literature (19). This comparison, as shown in Figure 8 for the center deflection indicates very good agreement with a maximum difference of 1.4%.

Multicell Bridge Deck: The Multicell bridge deck model, which was investigated both experimentally and analytically by Sawko and Cope (16), was analysed using the proposed model. Concentrated loads were applied at different web locations along the mid-span axis. The results of the analyses are shown in Figure 9 together with the experimental results and analysis reported by Sawko and Cope (16). This comparison shows that the proposed model predictions are in good agreement with the experiment data and generally give better predications than the other analyses.

Convergence Study for Thin Plates: A convergence study for a thin square plate under uniformly distributed loading was carried out for simply supported edges with shear restraint along the edges. A thickness to span ratio of 0.01 was examined and deflections, bending moments, and shear forces were compared with theoretical thin plate solutions in Table 1 and Figure 10. The convergence is fairly rapid in all cases.

Nonlinear Analysis of a Thin Plate: A simply supported square plate subjected to uniformly distributed load was analysed using the proposed model both in linear and nonlinear stages. The plate dimensions, material properties and the finite element mesh used for one quarter of the plate are given in Figure 11(a). These are the same as those available in the literature (17). Three layers were assumed through the plate thickness to reasonably represent the three integration points through the thickness considered in Reference 17. Figure 11(b) shows the load versus center deflection results from the proposed model (Analysis I) together with the reported solution. The considerable vertical shift between the two analyses is attributed to the fact that the proposed model

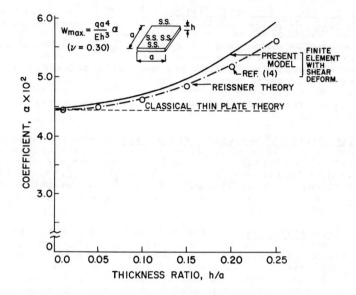

Figure 7. Isotropic Single Layer Plates.

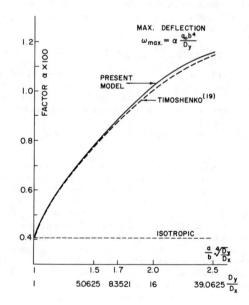

Figure 8. Orthotropic Single Layer Plates.

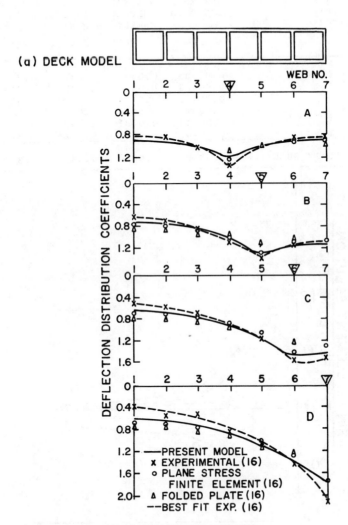

Figure 9. Multicell Bridge Deck Model.

TABLE 1. SIMPLY SUPPORTED SQUARE PLATE UNDER UNIFORM LOADING

MESH IN A SYMMETRIC QUARTER	CENTRAL DEFLECTION $\times q\frac{L^4}{D}$	CENTRAL BENDING MOMENT $\times qL^2$	CORNER TWISTING MOMENT $\times qL^2$	MID-EDGE FORCE $\times qL$
1 × 1	0.00506	0.0546	0.0319	0.340
2 × 2	0.00433	0.0497	0.0329	0.313
3 × 3	0.00418	0.0487	0.0328	0.324
4 × 4	0.00413	0.0483	0.0327	0.326
5 × 5	0.00410	0.0481	0.0327	0.328
6 × 6	0.00409	0.0481	0.0326	0.329
THIN PLATE SOLUTION	0.00406	0.0479	0.0325	0.329

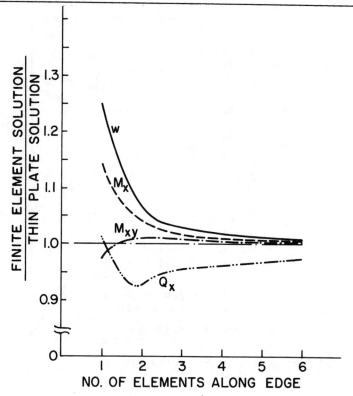

Figure 10. Convergence Study for Thin Plates Under Uniformly Distributed Load.

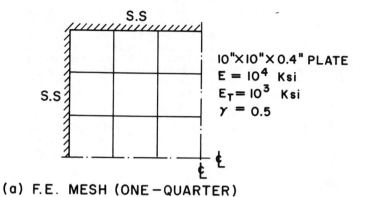

(a) F.E. MESH (ONE-QUARTER)

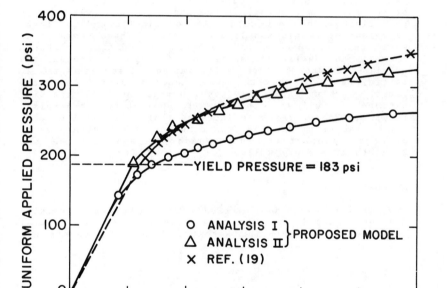

(b) LOAD-CENTRAL DEFLECTION

Figure 11. Nonlinear Analysis of Uniformly Loaded Simply Supported Plate.

considers the stresses at the extreme fibres for checking yielding. However, the other analysis (17) considered stresses at integration points located within the plate thickness. To check the validity of this argument, another analysis (Analysis II) was performed using the proposed model but using stresses at the same locations as those used in Reference 17. As shown in Figure 11(b), these predictions confirmed the previously mentioned reasoning.

It should be noted that an analysis of this plate was also performed using a single layer and that the results were very similar to those for the three layer analysis.

CONCLUSIONS

Details of the development of the proposed macroscopic finite element model for masonry have been presented. The accuracy of the model has been verified by comparison with known solutions for elastic and nonlinear behaviour of single and layered plates. These comparisons have been included in this paper. In addition, although there is not sufficient space in this paper to present the details of the failure criteria and associated tests, initial analyses show that this model accurately predicts first cracking and the develop ment of the cracking pattern for full scale block walls subject to out-of-plane bending. Therefore it is intended that this model be used for parametric studies to document the influence of material properties and boundary conditions on the strength of masonry walls.

ACKNOWLEDGEMENTS

This research was carried out at McMaster University and was funded by Operating Grants from the Natural Sciences and Engineering Research Council of Canada and the Masonry Research Foundation of Canada.

REFERENCES

1. Allen, H.G., "Analysis and Design of Structural Panels". Pergamon Press, Oxford, 1969.
2. Arya, S.K. and Hegemier, G.A., "Finite Element Method for Interface Problems", Journal of Struc. Div., ASDE. Vol. 108, No. ST2, Feb. 1982, pp. 327-342.
3. Balachandran, K., "An Investigation of the Strength of Concrete Masonry Shear Wall Structures", Ph.D. Thesis, University of Florida, 1974.
4. Basu, A.K. and Dawson, J.M., "Orthotropic Sandwich Plates -- Part 1: Dynamic Relaxation Treatment, Part 2: Analysis and application to Multicell and Voided Bridge Decks", Proc. of Instn. of Civil Engrs., 1970, Suppl., pp. 87-115.
5. Drysdale, R.G. and Essawy, A.S., "Out-of-Plane Bending of Concrete Block Walls", presented at Oct. 1984, ASCE Convention, San Francisco, Clif. Accepted for publication in the Journal of Struc. Div., ASCE, 14 pages.
6. Essawy, A.S. and Drysdale, R.G., "Capacity of Block Masonry Under Uniformly Distributed Loading Normal to the Surface of the of the Wall", Proc. of 3rd. Canadian Masonry Symposium, June 1983, Paper No. 39, 14 pages.
7. Gazzola, E.A., Drysdale, R.G. and Essawy, A.S., "Bending of Concrete Masonry Wallettes At Different Angles to the Bed Joints", Proc. of 3rd. North American Masonry Conf., June 1985, Paper No. 27, 14 Pages.
8. Goodman R.E. and St. John, C., "Finite Element Analysis for Discontinuous Rocks", Numerical Methods in Geotechnical Eng., pp. 148-175. Also, "Deep Foundation", pp. 239-245.
9. Hinton, E., and Owen, D.R.J., "Finite Element Programming", Academic Press, London, 1977.
10. Hinton, E., Razzaque, A., Zienkiewicz, O.C., and Davis, J.D., "A Simple Finite Element Solution for Plates of Homogeneous, Sandwich and Cellular Construction", Proc. of Instn. of Civil Engrs., Vol. 59, Part 2, March 1975, pp. 43-65.
11. Mindlin, R.D., "Influence of Rotatory Inertia and Shear on the Flexural Motions of Isotropic, Elastic Plates", Journal of Applied Mech. Div., ASME, Vol. 18, March 1951, pp. 31-38.
12. Page, A.W., "Finite Element Model for Masonry", Journal of Struc. Div., ASCE, Vol. 104, No. ST8, Aug. 1978, pp. 1267-1285.
13. Page, A.W., "A Biaxial Failure Criterion for Brick Masonry in the Tension -- Tension Range", Intern. J. of Masonry Construc., Vol. 1, No. 1, March 1980, pp. 26-29.

14. Pryor, Jr., C., "Finite Element Analysis of Laminated Anisotropic Plates Including Transverse Shear Deformations", Ph.D. Thesis, Virginia Polytechnic Institute, 1970, 130 pages.
15. Pryor, Jr., C.W. and Barker, R.M., "A Finite-Element Analysis Including Transverse Shear Effects for Applications to Laminated Plates", AIAA Journal, Vol. 9, No. 5, May 1971, pp. 912-917.
16. Sawko, F. and Cope, R.J., "Analysis of Multi-Cell Bridges Without Transverse Diaphragms: A Finite Element Approach", the Structural Engineer, Vol. 47, No. 11, Nov. 1969, pp. 455-460.
17. Shehata, A.A., "Finite Element Modelling and Elasto-Plastic Analysis of RHS Double Chord T-Joints", Ph.D. Thesis, McMaster University, 1983.
18. Soliman, M.I., "Behaviour and Analysis of a Reinforced Concrete Box Girder Bridge", Ph.D. Thesis, McGill University, Montreal, Canada, 1979.
19. Timoshenko, S.P. and Krieger, W., "Theory of Plates and Shells", Second Edition, McGraw Hill Book, New York, 1959.
20. Zienkiewics, O.C., "The Finite Element Method", McGraw-Hill Book, London, 1977.

APPENDIX I

NOTATIONS

E_x and E_y	=	the modulii of elasticity in x and y directions, respectively.
E_w	=	the modulus of elasticity for the web material.
G_f	=	the shear modulus of the face shell material.
G_{xy}, G_{xz} and G_{yz}	=	the shear modulii in x-y, x-z and y-z planes, respectively.
G_w	=	the shear modulus of the web material.
h	=	the height of the Cellular plate measured between center lines of the face shells (see Figure 4).
L	=	the transverse span of the panel.
n	=	the number of layers in the multilayer plate.
N_i, N_i^o and N_i^s	=	the shape functions for the 12 DOF nonconforming plate bending, 8 DOF plane stress and 8 DOF transverse shear rectangular elements, respectively.
t_f and t_w	=	the thickness of the face shell and the web of the cellular plate, respectively.
u, v and w	=	the displacements in the coordinate directions x, y and z, respectively.
u^o and v^o	=	the middle surface displacements in the coordinate directions x and y, respectively.
x, y and z	=	the cartisian coordinate directions.
α	=	the factor introduced to account for non-uniform shear deformations.

α_c and β_c	= the reduction factors for the modulii of elasticity and the shear modulii, respectively, for cracked elements.
γ_{xy}, γ_{xz} and γ_{yz}	= the shear strains in x-y, x-z and y-z planes, respectively.
γ_{xy}^o	= the shear strain in the xy plane (the middle surface).
ε_x and ε_y	= the strains in x and y directions, respectively.
ε_x^o and ε_y^o	= the middle surface strains in the x and y directions, respectively.
ξ and η	= the nondimensional coordinates with the origin at the left bottom corner.
θ_x and θ_y	= the total section rotations about the y and x axis respectively.
λ	= the cross web spacing.
μ	= the coefficient of friction between mortar and unit.
ν_{xy} and ν_{yx}	= Poisson's ratios of an orthotropic material in the y and x directions, respectively.
ν_w	= the Poisson's ratio of the web material.
σ_c	= the compressive stress normal to the crack.
σ_{so}	= the cohesion or the shear bond strength at zero precompression.
σ_x, σ_y and σ_z	= the normal stresses in the x, y and z directions, respectively.
τ_{xy}, τ_{xz} and τ_{yz}	= the shear stress in the x-y, x-z and y-z planes, respectively.
ϕ_x and ϕ_y	= the uniform shear deformations in x-z and y-z planes, respectively.

MASONRY WALLS MODEL

$[\bar{B}]_{8\times 28}$ = the generalized strain matrix relating the generalized strain vector to the displacement vector.

$[D]^k_{5\times 5}$ = the elasticity matrix for layer k.

$[D_{cr}]_{5\times 5}$ = the modified elasticity matrix for the cracked element.

$[D_{ef}]$ = the extension/flexure elasticity submatrix.

$[D_s]_{2\times 2}$ = the shear elasticity submatrix.

$[\bar{D}]_{8\times 8}$ = the total rigidity matrix.

$[\bar{D}_c]_{3\times 3}$, $[\bar{D}_e]_{3\times 3}$, $[\bar{D}_f]_{3\times 3}$ and $[\bar{D}_s]_{2\times 2}$ = the extension/flexure coupling, extension, flexure, and shear rigidity submatrices.

$[K_e]_{28\times 28}$ = the element stiffness matrix.

$\{M\}, \{N\}$ and $\{Q\}$ = the internal bending moment, in-plane force and shear force vectors, respectively.

$\{U\}$ = the displacement vector.

$\{\varepsilon\}, \{\bar{\varepsilon}\}$ and $\{d\varepsilon\}$ = the strain, generalized strain and incremental strain vectors, respectively.

$\{\varepsilon_s\}$ = the shear strain vector.

$\{\sigma\}, \{\bar{\sigma}\}$ and $\{d\sigma\}$ = the stress, generalized stress and incremental stress vectors, respectively.

$\{\chi\}$ = the vector containing the curvatures including the shear deformations.

BEHAVIOR OF CAVITY MASONRY WALLS UNDER
OUT-OF-PLANE LATERAL LOADING: AN ANALYTICAL APPROACH

Ahmad A. Hamid[1], M. ASCE, and Fred A. Carruolo[2]

ABSTRACT

The current North American masonry codes which address the lateral load distribution in cavity walls assumes that each wythe carries lateral load in proportion to its flexural rigidity regardless of the type of boundary. It is also assumed that the load transfer between the two wythes via metal ties is uniform. In this paper, a finite element model is used to analyze cavity walls in an attempt to understand their complex behavior and to evaluate the adequacy of the assumptions embodied in the current simplified design approach. The model accounts for material orthotropy and considers different geometric properties and boundary conditions to study wall deflections, tie forces and wall moments and stresses. The results show that boundary conditions of the two wythes have a significant effect on wall behavior and that load transfer and tie forces are not uniform. It is recommended that the current design approach should be modified to account for different boundary conditions and nonuniformity of load transfer between the two wythes.

INTRODUCTION

Masonry cavity wall is a type of masonry wall construction in which a continuous air space, from 2 to 4 1/2 in. (51 to 114 mm) thick, is provided inside the wall. The majority of cavity walls have an outer wythe of brick and an inner wythe of concrete block which may or may not be a load bearing. Such walls, if properly designed and constructed have excellent moisture and thermal resistance (8).

The current American masonry codes (1,6,7) which address the lateral load distribution in cavity walls assume that each wythe carries lateral load in proportion to its flexural rigidity, EI, regardless of the type of boundary. It is also assumed that the load transfer between

---

[1,2] Associate Professor and Graduate Student, respectively, Department of Civil Engineering, Drexel University, Philadelphia, PA.

the two wythes via metal ties is uniform. No documented experimental data is available in the literature to help understanding the behavior of cavity masonry walls. Brown and Elling (2) developed a simple discrete mathematical model to analyze two masonry wythes interconnected by discrete springs using a displacement method. In this paper, a finite element model is used to analyze cavity walls in an attempt to better understand their complex behavior and to evaluate the adequacy of the assumptions embodied in the current simplified design approach. The effects of different boundary conditions on wall deflections, tie forces, load distribution and wall moments and stress are studied.

ANALYTICAL MODEL

It is the objective of this study to investigate the effect of different geometric parameters and boundary conditions on the out-of-plane behavior of cavity in the precracking stage under service loads. Therefore, an elastic finite element model was developed using ANSYS which is a general purpose finite element program (3). A typical finite element mesh is shown in Fig. 1. Orthotropic plate bending elements (rectangular shells) are used for the two wythes and a three dimensional spare element is used for the metal ties. A total of 86 elements and 72 nodes are used.

Four different boundary conditions, I, II, III and IV are considered which represent the most common boundary conditions for exterior cavity walls with different construction details. Schematics of the four boundary conditions are presented in Fig. 2. Boundary condition IV represents a cavity wall with a shelf angle detail where the top of the brick wythe is free because of the compressible joint underneath the shelf angle (4).

Material properties used in the model are listed in Table 1. The ANSYS orthotropic finite elements allow uncoupled properties in the two orthogonal directions to be input directly to the material constant matrix. Different moduli of elasticity normal and parallel to bed joints, as shown in Table 1, are considered to adequately account for material anisotropy.

RESULTS

A 10 ft by 10 ft (3.05m x 3.05m) cavity wall is analyzed under a uniformly distributed lateral wind load of 40 psf (1.9 KPa). The output of the program contains displacements and stresses at the two faces of the block and brick wythes, moments at critical sections of the wythes and tie forces and displacements. Table 2 summarizes the results for different cases studied. Stresses and deflections are calculated at the middle surface of each wythe.

DISCUSSION OF RESULTS

Wall Deflections - Comparing deflections for different boundary conditons indicates the significant effect of boundary conditions on wall deflections and consequently on the lateral load distribution between the two wythes as can be seen in Table 2. This is attributed to the fact that boundary conditions affect member stiffness (10). Com-

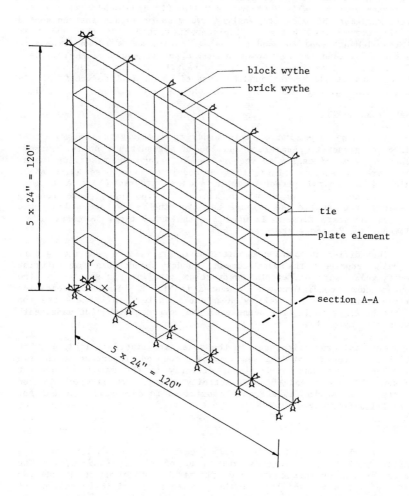

Fig. 1 - Finite Element Idealization of Cavity Walls

Table 1 - Material Properties

	Brick Wythe[3]	Block Wythe[9]	Metal Ties
E_x, ksi (GPa)[a]	3500 (24.15)	950 (6.56)	29000 (200.1)
E_y, ksi (GPa)[b]	6300 (43.47)	2040 (14.08)	-
E_y/E_x	1.8	2.15	-
Poisson's ratio	0.2 (ν_{xy})	0.2 (ν_{xy})	0.3 (ν_z)

a - Modulus of Elasticity in the x direction
b - Modulus of Elasticity in the y direction

paring results of Case 1 and Case 3, Table 2, indicates that continuous wythes are stiffer and therefore carry more of the lateral load. It is obvious from these results that the current codes provisions (1,6,7) in ignoring the boundary conditions and in using flexural regidity, EI, as the sole criterion for lateral load distribution, are inaccurate. This simplification may result in an unconservative design for either the brick or the block wythe.

Figure 3. shows the deflected shapes, which are automatically plotted by the postprocessor, for Case 1 and Case 4. As can be seen, free boundary of the brick wythe, boundary condition IV, Fig. 1, results in large deflection compared to walls with supported boundary.

Wall Moments and Stresses - Results presented in Table 2 indicate that wall stiffness reflected by boundary conditions have a significant effect on wall moments and consequently on wall stresses. Stiffer boundaries cause less moments in both wythes. Inaccurate account of actual boundary conditions would lead to gross error in estimating the critical moments at each wythe, which could cause undesirable cracks under service loads.

Figure 4 shows a typical plot of stress contours of Case 1 and Case 4. Comparing the stress contours of the two wythes indicates different patterns which is attributed to difference in boundary condition and in type of loading applied at each wythe. For Case 1, the two wythes have similar stress pattern with different intensity, whereas for Case 4 different patterns are apparent. This indicates that the simple design approach assuming that the two wythes are subjected to uniform loads may not be accurate in case of dissimilar boundaries.

Effect of Relative Flexural Rigidity - The relative flexural rigidity, EI, of the two wythes is altered by using block back-up of different thicknesses; 4", 8" and 12" (102,203 and 305 mm) keeping the thickness of the brick wythe the same, 4" (102mm). Figure 5 shows the variation of the percentage of load carried by the outer brick wythe with the

STRUCTURAL MASONRY ANALYSIS

Table 2 – Summary of Results

Case	Wall Thickness in (mm)		Type of Boundary Condition (a)	Percent Load Sharing by Brick Wythe	Moment (b) lb-in (kN-mm)		Maximum Tie Force	
	Brick Wythe	Block Wythe			Brick Wythe	Block Wythe	Value, lb (N)	location (c) in (mm)
1	4 (102)	8 (203)	I	38.8	2920 (333)	8672 (976)	158 (700)	48,0 (1219,0)
2	4 (102)	8 (203)	II	29.8	1230 (138)	3510 (395)	155 (687)	48,72 (1219,1829)
3	4 (102)	8 (203)	III	61.2	1370 (154)	2240 (252)	104 (460)	48,72 (1219,1829)
4	4 (102)	8 (203)	IV	29.4	660 (74)	2420 (272)	309 (1369)	48,96 (1219,2438)
5	4 (102)	4 (102)	I	74.5	8620 (970)	2970 (334)	291 (1289)	48,0 (1219,0)
6	4 (102)	12 (305)	I	28.3	1090 (123)	10500 (1181)	157 (696)	48,48 (1219,1219)

a) Refer to Fig. 2

b) Maximum moment, M_x, at section A-A, Fig. 1

c) x and y coordinates, refer to Fig. 1

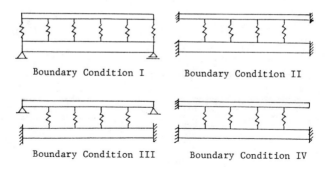

Fig. 2 - Types of Boundary Conditions

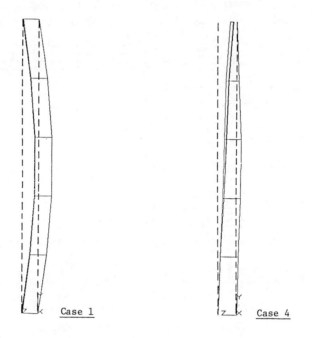

Fig. 3 - Deflected Shapes of Cases 1 and 4

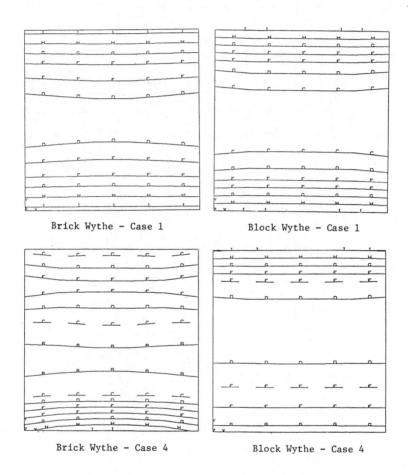

Fig. 4 - Stress Contours for Cases 1 and 4

CAVITY MASONRY WALLS BEHAVIOR

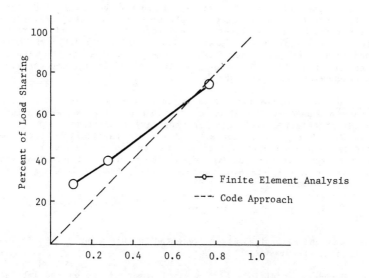

Fig. 5 - Effect of Relative Rigidity on Percent of Load Sharing by the Brick Wythe

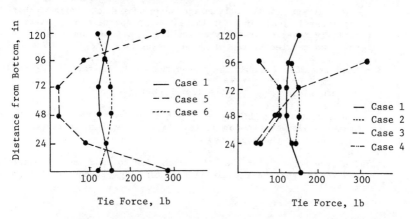

a) Effect of Relative Rigidity   b) Effect of Boundary Condition

Fig. 6 - Effect of Relative Rigidity and Boundary Conditions on Tie Forces

relative flexural rigidity for simply supported boundary conditions (Cases 1, 5 and 6). The results clearly indicate that the stiffer the brick wythe the higher the lateral load it carries. The load distribution to each wythe in proportional to its relative rigidity, which is recommended by different codes, would underestimate the load carried by the outer brick wythes. Therefore, factor of safety against cracking of the brick wythe will be reduced.

Tie Forces - Maximum tie forces for each case is presented in Table 2. Figure 6 shows the variation of tie forces along the wall height for different relative rigidities and boundary conditions. As can be shown in Fig. 6(a), tie forces are not uniform particularly for more flexible back-up. Large tie forces are created due to the free boundary of the brick wythe at top, see Fig. 6(b). The results indicate that the two wythes do not carry uniform loads as commonly assumed in the simplified design approach.

CONCLUSIONS

Based on the analytical study presented in this paper the following conclusions can be drawn:

1 - Boundary conditions have a significant effect on the lateral load distribution to each wythe of cavity masonry wall. Neglecting wall continuity may result in large errors in load distribution.

2 - The current simplified design approach which assumes that the lateral load is distributed in proportion to relative rigidity reveals inaccurate results that may underestimate the load carried by the block or the brick wythe.

3 - Load transfer between the two wythes and tie forces are not uniform. Providing ties at equal spacing results in a nonuniform factor of safety against buckling.

4 - The more the flexibility of the back-up wythe the higher the nonuniformity of tie forces is. High forces are attracted by the ties near the supports.

5 - The free boundary of the brick wythe due to compressible joint underneath the shelf angle results in large forces in the top ties.

RECOMMENDATIONS

It is recommended that current design approach based on relative rigidity for load distribution should be modified to account for different boundary conditions (i.e. to be based on relative stiffness rather than relative rigidity). It is also recommended to increase tie area requirements near floor levels in case of shelf angle detail and/or flexible back-up to achieve a more uniform factor of safety against buckling.

ACKNOWLEDGEMENTS

The writers would like to thank Mr. M. Cain of Drexel University Scientific Computation Center for his cooperation and advice on the PRIME system and on ANSYS program. The support of Swanson Analysis System Inc., Houston, Pennsylvania in providing ANSYS to Drexel University is gratefully acknowledged.

REFERENCES

1. American Concrete Institute Standard, ACI 531-79, "Building Code Requirement for Concrete Masonry Structures," Detroit, Michigan, 1979.

2. Brown, R. and Elling, R., "Lateral Load Distribution in Cavity Walls," Proceedings of the Fifth International Brick Masonry Conference, Washington, D.C., October 1979.

3. Desalvo, J. and Swanson, A., "ANSYS Engineering Analysis System User's Manual," Vols. 1 and 2, Swanson Analysis System Inc., Houston, Pennsylvania, 1983.

4. Hamid, A., Becica, I. and Harris, H., "Performance of Brick Masonry Veneer," Proceedings of the Seventh International Brick Masonry Conference, Melbourne, Australia, February 1985.

5. Hamid, A. and Drysdale, R., "Behavior of Brick Masonry Under Combined Shear and Compression Loading," Proceedings of the 2nd Canadian Masonry Symposium, Ottawa, Canada, June 1980.

6. International Conference of Building Officials, Uniform Building Code, 1976 Edition, Whittier, California.

7. National Concrete Masonry Association, Specifications for the Design and Construction of Load-Bearing Concrete Masonry, Herndon, Virginia, May 1976.

8. National Concrete Masonry Association, "Concrete Masonry Cavity Walls," NCMA TEK 62, Herndon, Virginia, 1978.

9. Nunn, R., "Planar Mechanics of Fully Grouted Concrete Masonry", Ph.D., Thesis, University of California, San Diego, 1980.

10. Timoshenko, Stephen, and Young, D. H., Elements of Strength of Materials, Fourth Edition, 1962, D. Van Nostrand Company, Inc., Princeton, New Jersey.

Shear Design of Masonry Walls

By Hans Rudolf Ganz [1] and Bruno Thürlimann, F. ASCE [2]

Abstract: The paper presents a summary of a research project carried out at the Swiss Federal Institute of Technology Zürich on the behaviour of masonry shear walls. One objective of this project is to develop a design method based on the ultimate strength of masonry walls. A general failure criterion for the biaxial compressive strength of masonry is presented. A statically admissible stress field consisting of "Discrete Compression Diagonals" is shown and lower bounds of the ultimate shear load are calculated using the theory of plasticity. The governing parameters of the interaction normal force-shear are emphasized. The extension to the case of an additional bending moment is shown and corresponding interaction curves are presented. The extension from the single shear wall to shear wall systems is indicated. Finally, the theoretical results are compared with full-scale tests on single shear walls.

Introduction

In the design of steel and concrete structures the theory of plasticity has been successfully used to determine the ultimate strength. Loads, derived on the basis of the static theorem, are preferred as they are lower bounds of the ultimate load. Furthermore, a statically admissible stress distribution in the whole structure is known such that a proper detailing of all components is possible. In order to apply the lower bound theorem, both failure criterion for masonry and statically admissible stress fields are needed.

The practical application of the theory of plasticity requires a sufficient amount of ductility of the material. Deformation-controlled experiments prove that masonry shear walls provide this ductility when moderate normal forces are applied (3).

This paper is restricted to the calculation of the ultimate strength of shear walls under static loading. The influence of cyclic loading is not included although some tests have been performed (3).

---

1) Research Engr., Institute of Structural Engineering,
   Swiss Federal Institute of Technology Zürich, Switzerland

2) Professor of Structural Engineering,
   Swiss Federal Institute of Technology Zürich, Switzerland

Failure Criterion for Masonry

A failure criterion for masonry has already been published (1). In the following, its derivation is briefly summarized.

Perforated bricks are composed of both biaxially and uniaxially stressed elements. As Fig. 1 shows, the elements of a brick make it geometrically anisotropic. In addition, the mortar joints have an influence on whether the elements are biaxially or uniaxially stressed. Furthermore, the joints reduce the brick strength to the masonry compressive strength.

For the shear resistance of the bed joints, a Coulomb failure criterion with a zero-tension cut-off is assumed. Usually, the perpend joints do not enter the failure criterion in the case of biaxial compression (1).

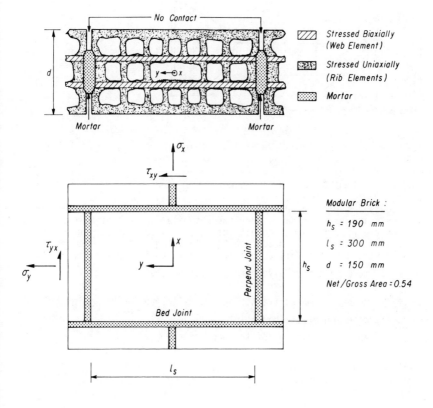

FIG. 1.- Masonry Element

Assuming a square failure criterion for the brick material and neglecting any tensile strength, the failure criterion for masonry is given by:

1: $\tau_{xy}^2 - \sigma_x \cdot \sigma_y = 0$ (no tension in bricks)

2: $\tau_{xy}^2 - (\sigma_x + f_{mx}) \cdot (\sigma_y + f_{my}) = 0$ (resistance of web and rib elements)

3: $\tau_{xy}^2 + \sigma_y \cdot (\sigma_y + f_{my}) = 0$ (resistance of web elements) (1)

4a: $\tau_{xy}^2 - (c - \sigma_x \cdot \tan\varphi)^2 = 0$ (sliding along bed joints)

4b: $\tau_{xy}^2 + \sigma_x \cdot [\sigma_x + 2 \cdot c \cdot \tan(\frac{\pi}{4} + \frac{\varphi}{2})] = 0$ (no tension in bed joints)

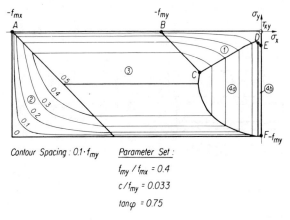

Contour Spacing: $0.1 \cdot f_{my}$

Parameter Set:
$f_{my} / f_{mx} = 0.4$
$c / f_{my} = 0.033$
$\tan\varphi = 0.75$

FIG. 2.- Failure Criterion for Masonry

Figure 2 shows a graphical representation of the failure criterion for perforated clay brick masonry for a typical set of parameters. Each of the five different conditions of the failure criterion represents a specific failure mode (see Eq. (1)). Neglecting the masonry tensile strength, the failure criterion includes four parameters:

f_{mx}: masonry compressive strength perpendicular to the bed joint

f_{my}: masonry compressive strength parallel to the bed joint

c : cohesion of the bed joint

φ : angle of friction of the bed joint

Stress Fields

Figure 3 shows a statically admissible stress field, referred to as Non-Centred Fan. The analysis of this and other stress fields is given in (1) and (4).

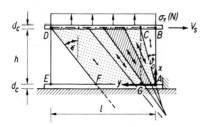

FIG. 3.- Non-Centred Fan.

The distributed normal force in Fig. 3 can be replaced by n equally spaced discrete loads N_i (Fig. 4). Each load $N_i < 0$ is carried by a discrete compression diagonal of width b_i. The inclined compression diagonals, $i = 1, 2, \ldots, n$, touch at the wall base, $x = 0$. This stress field is a fair approximation of the Non-Centred Fan.

In the following, a particular compression diagonal is considered. The forces acting on the diagonal can be expressed in terms of the discrete loads.

$$D_i = N_i \cdot \frac{1}{\cos\vartheta_i}$$

$$V_i = N_i \cdot \tan\vartheta_i \tag{2}$$

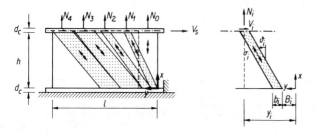

FIG. 4.- Discrete Compression Diagonals

The width b_i of the compression diagonal is a function of the inclination ϑ_i.

$$b_i = 2 \cdot [y_i - B_i - (h + \frac{d_c}{2}) \cdot \tan\vartheta_i] \leq \frac{l}{n} \quad ; \quad B_i = \sum_{k=1}^{i-1} b_k \tag{3}$$

The uniaxial principal stress σ_i in the compression diagonal can be determined by Eqs. (2) and (3). In addition, it should not violate the failure criterion for masonry.

$$\sigma_i = \frac{D_i}{d \cdot b_i \cdot \cos\vartheta_i} = \frac{N_i}{d \cdot b_i} \cdot \frac{1}{\cos^2\vartheta_i} \geqslant -f_m(\vartheta_i) \qquad (4)$$

Evaluating the failure criterion, Eq. (1), for uniaxial stress states, the uniaxial masonry compressive strength $f_m(\vartheta)$ is given by the minimum of the following expressions, see (1), (5):

$$\begin{aligned}
&2: \quad f_m(\vartheta) = f_{mx} & &, \vartheta \equiv 0 \\
&3: \quad f_m(\vartheta) = f_{my} & &, 0 < \vartheta \leqslant \frac{\pi}{2} \\
&4a: \quad f_m(\vartheta) = \frac{c}{\sin\vartheta \cdot \cos\vartheta - \cos^2\vartheta \cdot \tan\varphi} & &, \varphi < \vartheta \leqslant \frac{\pi}{4} + \frac{\varphi}{2} \\
&4b: \quad f_m(\vartheta) = 2 \cdot c \cdot \tan\left(\frac{\pi}{4} + \frac{\varphi}{2}\right) & &, \frac{\pi}{4} + \frac{\varphi}{2} \leqslant \vartheta < \frac{\pi}{2}
\end{aligned} \qquad (5)$$

Inserting the uniaxial compressive strength, Eq. (5), as well as the width b_i, Eq. (3), into Eq. (4), the inclination ϑ_i of the compression diagonal is obtained. If the conditions 3 and 4b of the failure criterion are governing, Eq. (4) can be solved directly.

$$\tan\vartheta_i = \kappa \cdot (h + \frac{d_c}{2}) + \sqrt{[\kappa \cdot (h + \frac{d_c}{2})]^2 - 1 - 2 \cdot \kappa \cdot (y_i - B_i)}$$

$$\begin{aligned}
&3: \quad \kappa = f_{my} \cdot \frac{d}{N_i} \\
&4b: \quad \kappa = 2 \cdot c \cdot \tan\left(\frac{\pi}{4} + \frac{\varphi}{2}\right) \cdot \frac{d}{N_i}
\end{aligned} \qquad (6)$$

However, if condition 4a is governing, Eq. (4) must be solved by iteration.

Thus, all unknown quantities are determined. Summation of the shear forces for all the inclined diagonals, i = 1,2,...n, establishes a lower bound V_s of the ultimate shear load V_u.

$$V_s = \sum_{i=1}^{n} V_i = \sum_{i=1}^{n} (N_i \cdot \tan\vartheta_i) \qquad (7)$$

The best lower bound V_s can be found by varying the number of vertical compression fields with subscript i=0. The superposition of the inclined diagonals with vertical fields does not affect the masonry strength $f_m(\vartheta)$, if the vertical stresses are limited to (see (1)):

$$\sigma_0 = \frac{N_0}{d \cdot b_0} \geqslant -(f_{mx} - f_{my}) \qquad (8)$$

Otherwise, condition 2 of the failure criterion becomes governing and $f_m(\vartheta)$ decreases and must be re-evaluated.

With increasing normal force N, the compression diagonals become wider until they touch under a constant inclination $\vartheta_i = \vartheta_1$. Thus, a parallel compression field is obtained (see (1), (4)). Any additional normal force is carried by vertical fields. Therefore, condition 2 must be checked for the superposition of inclined and vertical fields. Thus, a single stress field consisting of "Discrete Compression Diagonals" can be used for the total range of normal forces.

The calculated lower bounds depend on the distribution of the normal force. Figure 5 shows two possible distributions which establish almost the exact ultimate shear load. For arbitrary loading conditions, stiff concrete slabs are able to redistribute the normal force and thus increase the shear load to almost the exact ultimate load.

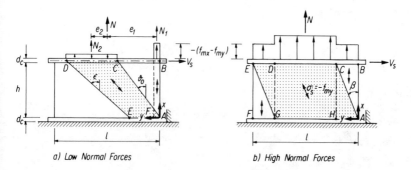

a) Low Normal Forces   b) High Normal Forces

FIG. 5.- Refined Stress Fields

Interaction Curves

Normal Force - Shear (N-V)

For a particular magnitude and distribution of normal force, a stress field can be evaluated and lower bounds of the N-V-interaction curve can be obtained. Figure 6 shows interaction curves for a non-uniformly distributed normal force.

Besides the distribution of the normal force, the interaction curves depend on the material parameters (f_{mx}, f_{my}, c and φ) and on the wall aspect ratio (l/h). As Fig. 6a shows, the relation shear load - aspect ratio is non-linear. For example, comparing the interaction curves for $l/h = 2$ and $l/h = 1$, the ratio of shear loads is $V_S(l/h=2)/V_S(l/h=1)$ = 3.0. The compressive strength f_{my} is the governing material parameter for shear wall failure. However, Fig. 6b shows that the shear load is less than proportional to the ratio f_{my}/f_{mx}. The influence of the parameters c and φ is limited to low normal forces.

Normal Force - Shear - Moment (N-V-M)

If normal and shear forces as well as a moment are applied to the wall, the actual stress distribution at the top of the wall is not known. Therefore, a statically admissible stress distribution must be chosen. Again, the stress field, composed of "Discrete Compression Diagonals", can be used to calculate lower bounds of the ultimate shear

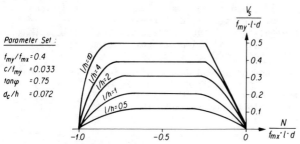

a) Influence of Wall Aspect Ratio l/h

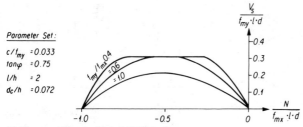

b) Influence of Masonry Strength Ratio f_{my}/f_{mx}

FIG. 6.- Normal Force - Shear Interaction Curves

load. In general, the discrete loads N_i will not be constant. However, N-V-M interaction surfaces can be established as for the case with constant loads.

For eccentric loads ($e_y > 0$) a simple method can be used to determine a lower bound of the ultimate shear load. By using a reduced wall length $\xi \cdot l$ a concentrically loaded shear wall can be obtained again as shown in Fig. 7.

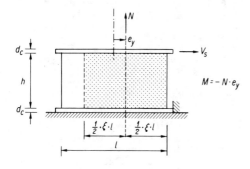

FIG. 7.- Shear Wall with Moment

$$\xi = 1 + 2\frac{M}{N \cdot l} \quad ; \quad 0 \geqslant \frac{N}{\xi \cdot l \cdot d} \geqslant -f_{mx} \tag{9}$$

Thus, the problem is reduced to the case without bending moment considered before. If all terms containing the wall length l are replaced by the reduced length $\xi \cdot l$, the same interaction curves (Fig. 6) can be used for cases with and without moment.

A simple relation exists between the ascending and the descending parts of the V-M-interaction curves. The moment at the base of the wall is a function of the moment at the top of the wall, the shear force and the wall height. The weight of the wall is neglected. Considering the sign convention for shear and moment (Fig. 7), the moment at the base must have a negative sign. If the moment is replaced by $N \cdot e_y$, the following relation can be established:

$$\frac{e_{y,des}}{l} = -(\frac{e_{y,asc}}{l} - \frac{V}{N} \cdot \frac{h}{l}) \tag{10}$$

The subscripts "des" and "asc" designate the descending and ascending parts of the curve. As shown in Fig. 8, there exists an auxiliary axis OC from which the points of the ascending and descending curves are equidistant for a particular value of shear.

Figure 8 shows V-M-interaction curves under constant normal forces. The evaluation of stress fields using a non-uniformly distributed normal force is compared with the reduced-wall-length method, Eq. (9). The axis OC is the same for both curves. The interaction curves are non-symmetric with respect to the eccentricity e_y. Thus, for relatively low normal forces the maximum shear resistance is obtained with negative moments (point C in Fig. 8). For higher normal forces, point C lies on a horizontal plateau. The curves have been calculated using the same material parameters as given in Fig. 2.

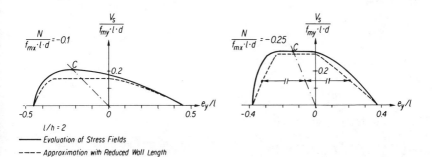

FIG. 8.- Shear - Bending Moment Interaction Curves

Influence of Reinforcement

Bed joints are planes of weakness in the masonry structure. Therefore, collapse mechanisms often develop along these joints. For such mechanisms there is no plastic strain along the bed joints and thus the reinforcement does no plastic work. Hence, bed-joint reinforcement produces no increase in ultimate shear load unless relatively high normal forces or rotational restraints change the governing collapse mechanism. A well anchored bed-joint reinforcement, however, can increase the ductility of shear walls considerably, e.g. (3).

The interaction curves for vertically reinforced walls may be obtained by translating the interaction curves for the unreinforced wall. The origin is moved within the yield surface of the reinforcement (linear combinations). In Fig. 9 this procedure is used to establish the N-V-interaction for a wall with a mechanical reinforcement ratio ω_x. Thus, all points with plastic strains $\varepsilon_n > 0$ on the interaction curve for the unreinforced wall are moved by ω_x in the direction of positive normal forces. In Fig. 9 the compressive strength of the reinforcement has been neglected. The same shear strength is obtained if the interaction of the unreinforced wall is used with a normal force N^*. N^* is the sum of the externally applied normal force N and yield load F_y of the vertical reinforcement.

$$N^* = N - F_y = N - \omega_x \cdot f_{mx} \cdot l \cdot d \quad ; \quad \omega_x = \frac{A_{sx}}{l \cdot d} \cdot \frac{f_y}{f_{mx}} \quad (11)$$

A_{sx} is the cross-sectional area and f_y the yield stress of the vertical reinforcement.

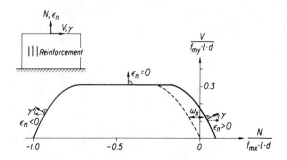

FIG. 9.- Influence of Vertical Reinforcement

Mild steel reinforcement has distinct disadvantages. The placing of the vertical bars is complicated as they must be threaded through the blocks and is therefore also time consuming. In addition, the bond between vertical bars and blocks is often questionable. Prestressed tendons (bonded or unbonded) concentrated at the wall edges do not have these disadvantages. Prestressed shear walls may be calculated again from the sum of the externally applied normal force and the prestressing forces

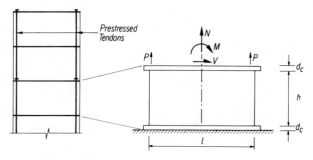

a) Structure   b) Loading of a Particular Shear Wall

FIG. 10.- Prestressed Shear Walls

$$N^* = N + 2 \cdot P \qquad (12)$$

where N and P have both negative signs. Thus, the strength of the shear wall is increased. In addition, reinforcement placed at the wall edges is more effective under moments. Finally, the potential cracking zones are prestressed and thus the behaviour under service loads will be improved, e.g. (3).

Plane Shear Wall Systems

In the following, the extension from a single shear wall to shear wall systems will be presented. As stated before, only static loading will be considered.

Solid Shear Walls

Figure 11 shows a multi-storey building under uniformly distributed horizontal and vertical loads. The shear load will be carried by a load-bearing system similar to a truss. The shear walls correspond to compression diagonals whereas the concrete slabs act as stiff ties. The slabs deflect the diagonal compression forces and distribute the shear load again to the rear parts of the wall. The inclination of the diagonals with respect to the vertical axis will decrease with increasing bending moment. This load-bearing system may be investigated with stress fields, i.e. with "Discrete Compression Diagonals". However, the computational effort involved is relatively large. Therefore, a first approximation may be based on the comparison of the loading vector $S(x)$ with the strength $R(x)$ in each storey.

$$R(x) \geqslant S(x) \qquad (13)$$

For the given structure, the loading vector $S(x)$ is indicated in Fig.11. The strength $R(x)$ is given by the three-dimensional N-V-M-interaction. However, using the reduced-wall-length method $\xi \cdot l$, a two-parameter problem can be obtained. Substituting the loading $S(x)$ in Eq. (9), the

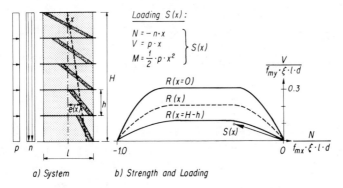

FIG. 11.- Solid Shear Walls

the reduced wall length is

$$\xi \cdot l = l \cdot (1 - \frac{p}{n} \cdot \frac{x}{l}) \tag{14}$$

Thus, the strength R(x) is obtained from N-V-interaction curves for different aspect ratios l/h, Fig. 11b.

The procedure given above remains valid for arbitrary structural systems. Therefore, the following remarks are limited to the determination of the strength R(x).

Coupled Shear Walls

The structural system to be considered is shown in Fig. 12. The strength R(x) of the system will be investigated for the walls S1 and S2. The shear-deflection curve of both walls is approximated by bilinear

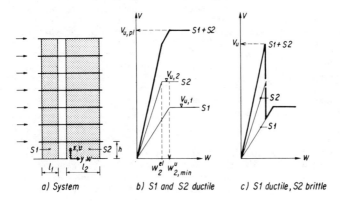

FIG. 12.- Coupled Shear Walls

curves. The determination of an appropriate wall stiffness will not be considered here. Due to different wall aspect ratios and different normal forces, shear wall S2 is stiffer than S1 and has a higher ultimate load. The response of the coupled walls is given by the sum of the two curves. If both walls are ductile enough the strength of the system is the plastic ultimate load (Fig. 12b).

$$V_{u,pl} = V_{u,1} + V_{u,2} \tag{15}$$

To reach the plastic load, shear wall S2 must have at least the following ductility:

$$\lambda \geqslant \frac{w_{2,min}^{u} - w_{2}^{el}}{w_{2}^{el}} \tag{16}$$

The meanings of the symbols are given in Fig. 12. Under relatively low normal forces $(N/(f_{mx} \cdot l \cdot d) \simeq -0.1)$ even unreinforced shear walls show ductilities $\lambda \simeq 5$ to 10, see (3). If the wall S2 fails in a brittle manner the ultimate shear load can be estimated according to Fig. 12c. Generally, the plastic ultimate load will be reached provided the wall dimensions and loading are not too different.

Shear Walls with Openings

Figure 13 shows two shear walls with openings. In cases of large openings, $c \simeq h$, the wall may be treated as divided along the vertical edges of the opening into two walls (Fig. 13a). The normal force N can be distributed in a statically equivalent manner.

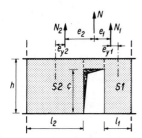
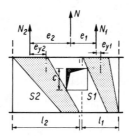

a) Openings $c \simeq h$ b) Openings $c \ll h$

FIG. 13.- Shear Walls with Openings

$$\begin{aligned} N_1 + N_2 &= N \\ N_1 \cdot e_1 &= N_2 \cdot e_2 \end{aligned} \tag{17}$$

The ultimate load of the walls S1 and S2 can now be calculated. Finally, the strength R(x) of a coupled system may be estimated as indicated above.

For small openings, c << h, the above mentioned procedure would provide conservative shear loads. Therefore, stress fields, e.g. the "Discrete Compression Diagonals", may be calculated which touch the perimeter of the opening (Fig. 13b). Again, a coupled system is obtained.

Restrained Shear Walls

Figure 14 shows a masonry pier framed by two stiff spandrels. These spandrels restrain the rotation of the pier ends. If the plastic rotation is totally prevented, $\kappa = 0$, the strength $R(x)$ corresponds to the maximum shear load in the V-M-interaction curve.

$$R(x) = V_{max} \qquad (18)$$

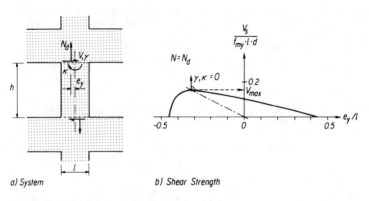

a) System   b) Shear Strength

FIG. 14.- Restrained Shear Walls

Comparison with Test Results

The comparison of the proposed failure criterion for masonry with test results has been given elsewhere (1), (5).

Seven full-scale tests on shear walls have been carried out (3). A test specimen is shown in Fig. 15. The flange widths of the specimens W4 and W5 were 900 mm. Bricks with a compressive strength of 37.4 N/mm^2 (based on gross area, area of perforations 46%, Fig. 1) and a cement mortar with a compressive strength of 28.5 N/mm^2 were used. The walls were loaded by either uniformly distributed normal forces or a concentrated normal force and a bending moment. The shear force was then applied through the upper slab and increased up to the failure of the wall (deformation-controlled). Table 1 gives a summary of the test programme as well as the experimental and theoretical ultimate shear loads.

SHEAR DESIGN OF MASONRY WALLS

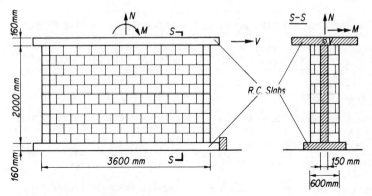

FIG. 15.- Test Specimen

The material parameters were

    Masonry strength perpendicular to the bed joint: f_{mx} = 8.25 N/mm²
    Masonry strength parallel to the bed joint: f_{my} = 3.0 N/mm²
    Cohesion of the bed joint: c = 0.06 N/mm²
    Angle of friction of the bed joint: $\tan\varphi$ = 0.81

Close agreement between theory and experiment has been obtained.

Test	N [kN]	M [mkN]	Bed-Joint Reinforcement [mm²/m]	Loading History	Experiment V_u [kN]	Theory V_s [kN]	$\dfrac{V_u}{V_s}$
W1	- 415	0	-	static	260	266	0.98
W2	-1287	0	-	static	454	477	0.95
W3	- 415	0	196.4 [1]	static	273	266	1.03
W4	- 423	355.3	-	static	187	158	1.18
W5	- 424	729.3	196.4 [1]	static	365	350	1.04
W6	- 418	0	-	cyclic	247 [3]/231 [4]	266/-	0.93
W7	-1290	0	-	cyclic	491 [3]/438 [4]	477/-	1.03

                                                1) Yield Stress f_y = 506 N/mm²
                                                2) Prestressed P_u = -310 kN
                                                                 $M(P_u)$ = -471.2 mkN
                                                3) 1st cycle
                                                4) 10th cycle

TABLE 1.- Shear Wall Tests

Conclusions

As the comparison in the preceeding section shows, the theory of plasticity can be used to predict the ultimate load of masonry shear walls reasonably well, provided an appropriate failure criterion is used. The proposed failure criterion includes the main parameters of masonry. It is therefore applicable to different types of bricks (clay bricks, concrete blocks, calcium silicate blocks, etc.) and to different types of mortar.

Although masonry elements behave relatively brittle (2), shear walls framed with concrete slabs show sufficient ductility to reach the plastic ultimate load. The ultimate load can be estimated reasonably well using stress fields. The field composed of "Discrete Compression Diagonals" can be used to determine the N-V-M-interaction surface.

A new code based on ultimate strength design is under preparation in Switzerland. The design of masonry shear walls is based on the theoretical background presented in this paper.

Appendix.- References

1. Ganz, H.R., "Mauerwerksscheiben unter Normalkraft und Schub", (Masonry Walls Loaded by Normal Force and Shear), Institut für Baustatik und Konstruktion, ETH Zürich (Thesis in preparation).
2. Ganz, H.R., and Thürlimann, B., "Versuche über die Festigkeit von zweiachsig beanspruchtem Mauerwerk", (Experimental Strength of Biaxially Loaded Masonry), Institut für Baustatik und Konstruktion, ETH Zürich, Versuchsbericht Nr. 7502-3, 1982. Birkhäuser Verlag Basel, Stuttgart, Boston.
3. Ganz, H.R., and Thürlimann, B., "Versuche an Mauerwerksscheiben unter Normalkraft und Querkraft", (Tests on Masonry Walls Loaded by Normal Force and Shear), Institut für Baustatik und Konstruktion, ETH Zürich, Versuchsbericht Nr. 7502-4, 1984. Birkhäuser Verlag Basel, Stuttgart, Boston.
4. Ganz, H.R., and Thürlimann, B., "Plastic Strength of Masonry Shear Walls", 7th International Brick Masonry Conference, Melbourne, February 1985.
5. Thürlimann, B., and Ganz, H.R., "Bruchbedingung für zweiachsig beanspruchtes Mauerwerk", (Failure Criterion for Biaxially Loaded Masonry), Institut für Baustatik und Konstruktion, ETH Zürich, Bericht Nr. 143, 1984. Birkhäuser Verlag Basel, Stuttgart, Boston.

Load-Bearing Capacity of Masonry Walls in a
Critical Buckling Condition taking into Account
the Restraining Effect of Floor Slabs

Prof. Dr.-Ing. W. Mann and Dipl.-Ing. E. Leicher

Summary

A building constructed with masonry walls and concrete floor slabs will behave as a frame. The loads on the slabs produce bending moments in the wall. At the same time, however, the walls are restrained in the slabs. Due to this effect, the buckling length of the wall is reduced, and the load capacity is considerably enlarged.

In the paper this very important effect is studied. The frame is calculated according to the theory of II. order. The buckling length, the bending moments, and the load capacity are determined for homogeneous materials as well as for materials without tensile strength. Finally simple formulas for practical use are developed.

1. Introduction

Masonry walls are mainly loaded by vertical forces. Therefore the buckling effect has to be considered. This effect depends on several influences. There are bending moments in the wall resulting, for instance, from wind loads or earth pressure, or from the floor slabs restrained in the walls, which produce eccentricities of the vertical forces. On the other hand, in many buildings the connection between walls and floor slabs is rigid, so that the walls can be regarded as restrained in the slabs. Due to this effect, the buckling length of the wall is reduced and the load capacity is increased. This is a very important point for dimensioning walls.

In order to use all advantages of masonry a lot of investigations have been made on this problem. Finally it has to be considered that in practice there is not enough time to use complicated theories: For instance it is not possible to calculate any wall exactly as a deformated frame. It is necessary to have simple formulas which describe the reality in an approximate and conservative way.

Prof. Dr.-Ing. W. Mann and Dipl.-Ing. E. Leicher, Technische Hochschule Darmstadt, D-6100 Darmstadt, W. Germany

In the following paper the frame system consisting of masonry walls and concrete floor slabs is studied according to the theory of II. order. The results are shown in formulas and diagrams. From these results, simple formulas for practical use are developed.

2. Symbols

$A = d \cdot b$	area of cross section of the wall
d, b	thickness d and width b of the wall
e	eccentricity of the normal force N in the wall
e_a	accidental eccentricity as part of e
c	distance of N to the edge
h, l	height of the wall, length of the slab
h_k	buckling length or effective height of the wall
$\bar{\lambda} = h_k / h$	slenderness of the wall
$(EJ)_B, (EJ)_M$	bending stiffness of the concrete slab and of the masonry wall
$R = \dfrac{h}{l} \cdot \dfrac{(EJ)_B}{(EJ)_M}$	relation of stiffness
$\eta = \eta(e; \bar{\lambda})$	reduction factor for load capacity
$\eta_e = \eta(e; \bar{\lambda} = 0)$	partial reduction factor, considering e and the type of stress distribution
$n = \dfrac{2N}{q_1 \cdot l}$	figure symbolizing the number of floors
$\alpha = \sqrt{\dfrac{N}{(EJ)_M}} = \dfrac{\pi}{h}\sqrt{\dfrac{N}{P_E}}$	figure symbolizing the degree of utilisation; P_E = Euler-load
q_i	load on the concrete slab i

3. Stress Distribution in the Wall Section

Different types of stress distribution are used in different countries, according to fig. 1. Linear stress distribution (a) without tension stresses is commonly used. This type is, for instance, still applied for masonry in Germany. But it is known from compression tests that it is very conservative. Therefore it was sometimes proposed to use parabolic distribution (b), which is more complicated. The diagrams in the CIB-Code are based on a part of a parabola with a small section of tensile stresses (c). Another type is the parabola-rectangle (d), which is normally used for concrete. This form is rather complicated.

Nearly the same results are obtained with the rectangular stress block (e), which is used for masonry in Great Britain and for unreinforced concrete in Germany. It is the simplest type for calculation. For this reason and because it is close to practice in concrete, the rectangular stress block was chosen for the first draft of the new ISO-Code for masonry. The differences between these types are shown in fig. 1 and 2.

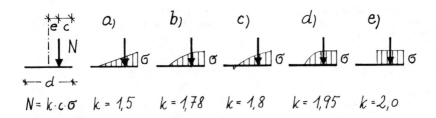

Fig. 1: Types of stress distribution

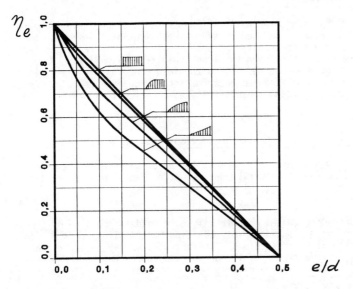

Fig. 2: Reduction factor η_e for the load capacity, depending on the eccentricity e, for the different types of stress distribution

For practical use it is important that the results are not only to take from diagrams, but can also be calculated by simple formulas. The maximum stresses σ for linear distribution (a) are:

$$\sigma = \frac{N}{A}\left(1 \pm \frac{6e}{d}\right) \quad \text{for } 0 \leqslant e \leqslant d/6 \tag{1a}$$

$$\sigma = \frac{2N}{3cb} \quad \text{for } \frac{d}{6} \leqslant e < \frac{d}{2} \tag{1b}$$

The rectangular stress block (e) can also be calculated in a simple manner. The maximum stress σ and the reduction factor η_e for the load capacity are:

$$\sigma = \frac{N}{2cb} \quad \text{for all } c = \frac{d}{2} - e \tag{2}$$

$$\eta_e = 1 - 2\frac{e}{d} \tag{3}$$

4. Buckling of a Hinged Wall with Constant Load-Eccentricity

4.1 Theoretical results

The buckling effect for walls without tensile strength considering theory of II. order is deduced in several publications. Some results for constant load-eccentricity are shown in fig. 3. The reduction factor η means the relation between the load capacity of a wall with an eccentricity e and a slenderness λ to the load capacity of a wall with $e = 0$ and $\lambda = 0$. This diagram does not include any accidental eccentricity. If an accidental eccentricity e_a occurs, it has to be added into e.

4.2 Simple approximation for the reduction factor η

In order to give a simple formula for practical use, it was proposed, see lit. (4), to describe the reduction factor η as a product

$$\eta = \eta_e \cdot \eta_\lambda \tag{4a}$$

where η_e, depending on the eccentricity e, considers the type of stress distribution and η_λ considers the reduction by the slenderness λ. In case of rectangular stress block, for instance, η_e may be taken from eq. (3). η_λ was developed by trying to find approaches to the curves in fig. 3. So the following equation was proposed:

$$\boxed{\eta = \left(1 - 2\frac{e}{d}\right) \cdot \frac{1}{1 + a\left(\frac{\lambda}{25}\right)^2}} \tag{4b}$$

$$a = 1 + 30\left(\frac{e}{d}\right)^2 \tag{4c}$$

LOAD-BEARING CAPACITY

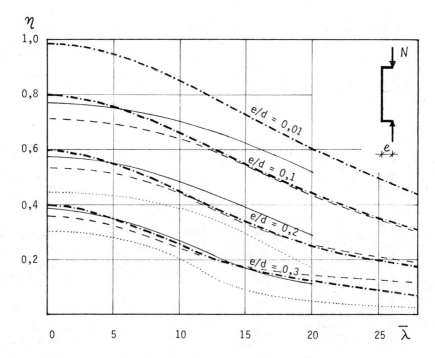

Fig. 3: Reduction factor η for different types of stress distribution

———— rectangular-parabola (d) according to lit. (1)

- - - - part of parabola (c) according to lit. (2), fig. 19, $\bar{\alpha} = 800$

········ linear (a) according to lit. (3), $\alpha_s = 400$, without e_a

—·—·— approximation according to chapter 4.2

This approximation is presented in fig. 3. One can see that it describes the character of η very well. η can be chosen more or less conservative by changing the factor a, if this is required.

Eq. (4b) does not include any accidental eccentricity e_a. If existing, e_a has to be added into e.

5. Walls and Floors as a Frame System

5.1 System

In many constructions walls and floors are connected in such a rigid manner that they act together as a frame. All kinds of frames can be calculated as a single storey frame by changing the bending stiffness of the slabs and the loads on the slabs. An example is shown in fig. 4. For this reason the following investigations have been made on a single storey frame.

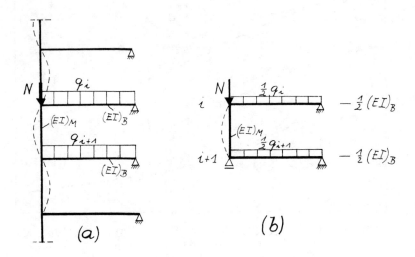

Fig. 4: Buckling frame systems
a) actual frame system;   b) substitute frame system

5.2 Bending Moments in the Frame System

The single storey substitute frame system was calculated exactly according to theory of II. order. The accidental eccentricity e_a is taken as being constant over the height of the wall. See fig. 5. Especially the bending moments M_1, M_1', M_2, M_2' and M_3 have been computed, in order to find the maximum value of these 5 moments in the wall. This maximum is here called max M (frame). It is decisive for dimensioning the wall.

LOAD-BEARING CAPACITY

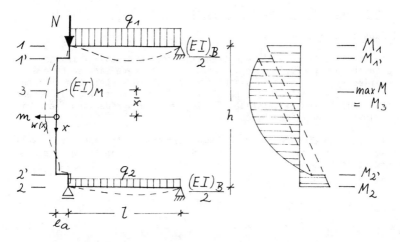

Fig. 5: Calculated frame system

The frame was calculated for homogeneous elastic material as well as for material without tensile strength. The results for homogeneous material are the following equations:

$$(EI)_M \cdot w(x)'' + N \cdot w(x) + N \cdot e_a + \tfrac{1}{2}(M_1+M_2) - (M_1-M_2) \cdot \tfrac{x}{h} = 0 \qquad (5)$$

$$M(x) = \left(N \cdot e_a + \tfrac{M_1+M_2}{2}\right) \cdot \tfrac{\cos \alpha x}{\cos(\alpha h/2)} - \tfrac{1}{2}(M_1-M_2) \cdot \tfrac{\sin \alpha x}{\sin(\alpha h/2)} \qquad (6)$$

$$\alpha = \sqrt{\tfrac{N}{(EI)_M}} = \tfrac{\pi}{h} \cdot \sqrt{\tfrac{N}{P_E}} \; ; \qquad P_E = \tfrac{\pi \cdot (EI)_M}{h^2} \qquad (7a)$$

$$R = \tfrac{h}{l} \cdot \tfrac{(EI)_B}{(EI)_M} \qquad (7b)$$

$$\tfrac{1}{2} \cdot (M_1+M_2) = \left[(q_1-q_2) \cdot \tfrac{l^2}{32} + e_a \cdot N\right] \cdot \tfrac{1}{1+\tfrac{3}{2\alpha h} \cdot R \cdot tg\tfrac{\alpha h}{2}} - e_a N \qquad (8)$$

$$\tfrac{1}{2} \cdot (M_1-M_2) = (q_1-q_2) \cdot \tfrac{l^2}{32} \cdot \tfrac{1}{1+\tfrac{3}{(\alpha h)^2} \cdot R - \tfrac{3}{2\alpha h} \cdot R \cdot \tfrac{1}{tg(\alpha h/2)}} \qquad (9)$$

$$M_3 = \left(N \cdot e_a + \tfrac{M_1+M_2}{2}\right) \cdot \tfrac{\cos \alpha \bar{x}}{\cos(\alpha h/2)} - \tfrac{M_1-M_2}{2} \cdot \tfrac{\sin \alpha \bar{x}}{\sin(\alpha h/2)} \qquad (10)$$

$$\sin \alpha \bar{x} = \tfrac{\tfrac{1}{2}(M_1-M_2)}{\sqrt{\tfrac{1}{4}(M_1-M_2)^2 + (N \cdot e_a + \tfrac{M_1+M_2}{2})^2 \cdot tg^2(\alpha h/2)}} \qquad (11)$$

6. Substitute Wall as Simplified System for the Buckling Effect

6.1 Substitute System

In practice there will be not enough time to calculate frames according to theory of II. order. Therefore it is necessary to have a simplified system, which simulates the real frame system. For this purpose it is convenient to use the hinged wall with constant load eccentricity; this system is described in chapter 4, its reduction factor is shown in fig. 3 or given by eq. (4b).

In order to compare the real frame system with the substitute wall system, it is necessary to know the bending moments in the wall system according to fig. 6.

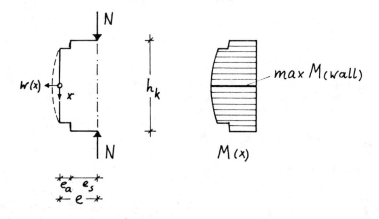

Fig. 6: Substitute wall

This wall is described by the following equations:

$$(EI)_M \cdot w(x)'' + N \cdot w(x) + N \cdot (e_a + e_s) = 0 \qquad (12)$$

$$M(x) = N \cdot (e_a + e_s) \cdot \frac{\cos \alpha x}{\cos(\alpha h_k/2)} \qquad \alpha \text{ see eq. (7)} \qquad (13)$$

$$\max M(wall) = N \cdot \frac{e_a + e_s}{\cos(\alpha h_k/2)} \qquad (14)$$

The values of h_k and e_s must be determined in such a way that the load capacity of the substitute wall is the same as that of the frame system.

6.2 Buckling Length or Effective Height of the Wall

In order to determine the buckling length h_k of the wall, it is convenient to compare the frame fig. 5 under the special condition $q_1 = q_2 = 0$ with the substitute wall fig. 6 under the special condition $e_s = 0$. The load capacity of these two systems is the same, if the bending moments max M are the same. Firstly the frame system:

In the case of $q_1 = q_2 = 0$ eq. (8), (9) and (10) simplify:

$$\bar{x} = 0; \quad M_1 = M_2; \quad \max M (frame) = M_3 = \frac{N \cdot e_a}{\cos(\alpha h/2) + \frac{3}{2\alpha h} \cdot R \cdot \sin(\alpha h/2)}$$

Now the substitue wall with $e_s = 0$: From eq. (14) follows

$$\max M (wall) = \frac{N \cdot e_a}{\cos(\alpha h_k/2)}$$

The values of max M are equal, if the denominators are equal:

$$\cos(\alpha h/2) + \frac{3}{2\alpha h} \cdot R \cdot \sin(\alpha h/2) = \cos(\alpha h_k/2) \qquad (15)$$

In the critical case of buckling eq. (7a) can be simplified:

$$N = N_k = \frac{\pi \cdot (EI)_M}{h_k^2}; \qquad \alpha h = \pi \cdot \frac{h}{h_k}$$

This value introduced into eq. (15) leads to the equation for h_k:

$$\frac{h_k}{h} \cdot tg\left(\frac{\pi}{2} \cdot \frac{h}{h_k}\right) = -\frac{2}{3} \cdot \frac{\pi}{R} \qquad (16)$$

The relation h_k / h according to eq. (16) is shown in fig. 7. This figure is taken from lit. (5). The dotted curves show the results for materials without tensile strength. In this case the buckling length of the wall is smaller than in the case of elastic material. The cracks reduce the wall stiffness and increase by this effect the degree of restraint in the slab. See also lit. (7) and (8).

As one can see in fig. 7, in nearly all practical cases it is sufficient to take $h_k = 0,75 \, h$. It may become somewhat higher, if the stiffness of the masonry wall is very large, which leads to small values for R. But in this case of stiff walls, the reduction of the buckling length is not important. In the new German Masonry Code DIN 1053 part 2 this effect is written in the follwing way:

external walls: $\quad h_k = h \, (1 - 0,15 \, R) \geqslant 0,75 \, h \qquad (17a)$

internal walls: $\quad h_k = h \cdot [1 - 0,15 \, R \cdot h \cdot (\frac{1}{l_l} + \frac{1}{l_r})] \geqslant 0,75 \, h \qquad (17b)$

In order to better understand the mechanical contents of fig. 7, see fig. 8. The frame wall acts between the hinged wall Euler-case 2 with $h_k = h$ and the restrained wall Euler-case 4 with $h_k = 0,5\,h$. The value $h_k = 0,75\,h$ is exactly between these two cases.

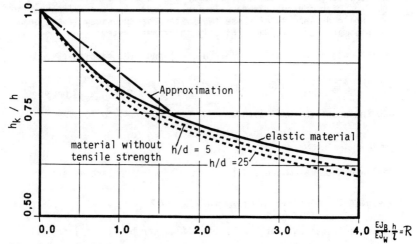

Fig. 7: Buckling length of a wall in a frame system
　——— elastic material, eq. (16)
　----- material without tensile strength
　—·—·— approximation, eq. (17a)

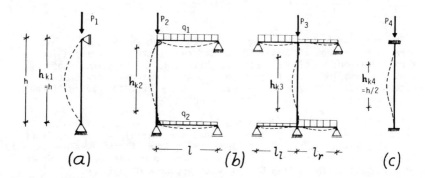

Fig. 8: Buckling systems
　　(a) Euler-case 2; (b) frame systems; (c) Euler-case 4

It is emphasized that this effect is proved by tests in some extend, see lit. (6).

It is obvious that the reduction of the effective height is only valid, if the connection between wall and concrete slab is rigid.

6.3 Substitute Wall Eccentricity

The usual calculation of the frame according to theory of I. order leads to the bending moments M_1 and M_2, resulting from the restraint of the concrete floor slabs. See fig. 9. From these, the moment M_m in the middle of the wall can be derived as $M_m = \frac{1}{2}(M_1 + M_2)$. In special cases, additional moments are to be added at this point, e. g. moments resulting from wind load or from earth pressure.

As the buckling of the wall is mostly influenced by the eccentricity e_m in the middle of the wall, it is easy to understand that the substitute wall system will closely approximate to the actual system, if $e_s = e_m$.

It is proved in chapter 6.5 that this equation is on the safe side.

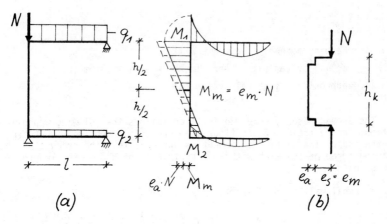

Fig. 9: Actual system (a) and substitute wall system (b) for the buckling effect

6.4 Simplified Method to calculate the Framework

The practical calculation of the frame wall has to be carried out in several steps.

Step 1: Bending moments M_1, M_2 and M_m in the frame wall, see fig. 10. These moments, resulting from theory of I. order, can be found either by an exact frame calculation or by approximation.

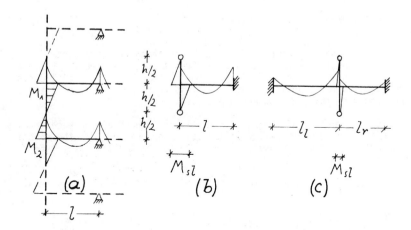

Fig. 10: Bending moments in a frame wall
actual system (a); approximate external system (b);
approximate internal system (c)

The German Masonry Code, for instance, allows the following approximation: The restraint of the floor slab produces a bending moment M_{sl} at the end of the slab: This moment is approximately the product of the bearing force of the slab with an eccentricity of 5 % of the length of the slab:

external walls: $\quad M_{sl} \approx \frac{1}{2} q \cdot l \cdot 0{,}05\, l \hfill (18a)$

internal walls: $\quad M_{sl} \approx \frac{1}{2} q_l \cdot l_l \cdot 0{,}05\, l_l - \frac{1}{2} q_r \cdot l_r \cdot 0{,}05\, l_r \hfill (18b)$

At the top floor, M_s equals the moment at the wall top:

$$M_1 \cong M_{sl}$$

At the intermediate floors, one half of M_{sl} goes down, the other half goes up the wall:

$$M_1 \cong M_2 = \frac{1}{2} \cdot M_{sl}$$

This approximation is simple, but of course not very exact. Other simplified formulas, which are more exact, are, for instance, given in lit. (3). They base on the simplified systems (b) and (c) in fig. 10, where the bending moment is assumed to be Zero at half of the wall

LOAD-BEARING CAPACITY 83

height. Moreover, these such calculated moments, which are derived for homogeneous material, are reduced to half their value because of several reasons: The masonry wall has no tensile strength; therefore its stiffness is smaller than calculated; the reinforcement of the slab will increase the slab stiffness; finally the bending moments have the character of restraint moments, which are not necessary for equilibrium. In this manner the moments at the end of the slab are for

external walls $\quad M_{sl} = \frac{1}{2} \cdot \frac{q l^2}{12} \cdot \frac{1}{1 + \frac{1}{3} R} \quad (19a)$

internal walls $\quad M_{sl} = \frac{1}{2} \cdot \frac{1}{12} \cdot (q_l \cdot l_l^2 - q_r \cdot l_r^2) \cdot \frac{1}{1 + \frac{1}{3} R (1 + l_l/l_r)} \quad (19b)$

for $l_l \geq l_r$; R see eq (7b) with $l = l_l$

intermediate floors $\quad M_1 \sim M_2 = \frac{1}{2} M_{sl}$

M_m follows from M_1 and M_2; occasionally moments from horizontal forces have to be added.

Step 2: Load capacity at top and bottom of the wall. As there is no important influence from buckling, the reduction factor can be taken from eq. (3)

Top: $\quad \eta_1 = 1 - 2 \frac{e_1}{d}$

$e_1 = \frac{M_1}{N} \pm e_a$

Bottom: $\quad \eta_2 = 1 - 2 \frac{e_2}{d}$

$e_2 = \frac{M_2}{N} \pm e_a$

Step 3: Middle of the wall height: The reduction factor follows from eq. (4b)

$$\eta_m = \left(1 - 2 \frac{e_m}{d}\right) \cdot \frac{1}{1 + a \left(\frac{\overline{\lambda}}{25}\right)^2}$$

$$e_m = \frac{M_m}{N} \pm e_a$$

a according to eq. (4c); $\overline{\lambda} = h_k / d$ with h_k according to eq. (17a, b).

The wall is to be dimensioned using the smallest of the values η_1, η_2 and η_m.

6.5 Comparative Calculations

In order to prove the proposed simplified method, submitted in chapter 6.4, comparative examples have been computed, variing the following factors:

Relation of bending stiffness $R = \dfrac{h}{L} \cdot \dfrac{(EJ)_B}{(EJ)_M}$;

relation of loads on floor slabs q_2 / q_1;

load N on the wall, symbolized by the figure n;

relation between actual force N and Euler-load $P_E = \pi^2 \cdot \dfrac{(EJ)_M}{h^2}$ (corresponding $\alpha \cdot h$);

elastic material and material without tensile strength.

All these factors were variied to that extend, which occours in common buildings. In this manner some thousand examples were computed.

For the wall system the effective height h_k was taken according to eq. (17a), that means the chain-dotted approximation of figure 7. In all examples the values max M (frame) according to chapter 5.2 and max M (substitute) according to chapter 6.4 have been computed. Herein, max M (substitute) results from the maximum of η_1, η_2 and η_m, with M_1 and M_2 calculated from the exact frame system according to theory of I. order. The relation

$$k = \dfrac{\max M \text{ (frame)}}{\max M \text{ (substitute)}}$$

was formed. Some typical results are shown in fig. 11a and 11b. All other results are similar to these figures.

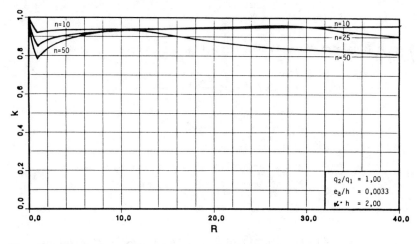

Fig. 11a: Typical examples for the relation k, showing the quality of the simplified calculation method

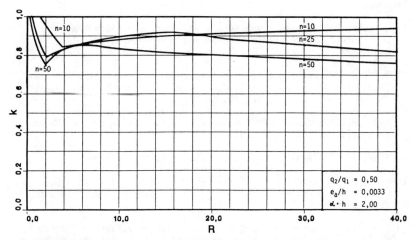

Fig. 11b: See description to fig. 11a

The relation factor $k = 1$ means that the bending moment and thereby the load capacity of the actual frame system is the same as that of the calculated simplified wall system. $k < 1$ means that the simplified system is on the safe side.

The results demonstrate that the proposed simplified method produces satisfactory concurrence with the actual system for all practical cases and that it is on the safe side.

Literature:

(1) Grasser, E.: Bemessung von Beton- und Stahlbetonbauteilen, Deutscher Ausschuß für Stahlbeton, Heft 220, 1979, page 121, fig. 4.27.

(2) Kukulski, W. and Lugez, J.: Resistance des murs en Béton non armé soumis a des charges verticales. Cahiers du Centre Scientifique et Technique du Batiment, Nr. 79, 4-1966.

(3) Mann, W.: Grundlagen der Bemessung von Ingenieurmauerwerk, Mauerwerk-Kalender 1985, page 39.

(4) Mann, W.: Explanation to ISO TC 179, SC 1, May 1985.

(5) Mann, W. and Leicher, E.: Investigations on the buckling effect in masonry walls restrained in the floor slabs, 7th IBMaC Melbourne 1985, page 643.

(6) Gremmel, M.: Zur Ermittlung der Tragfähigkeit schlanker Mauerwerkswände an Bauteilen in wirklicher Größe, Dissertation, Braunschweig 1978.

(7) Mann, W.: Sicherheitstheoretische Untersuchungen auf dem Gebiet des Mauerwerksbaues, Abschnitt 18: Abminderung der Knicklänge durch Deckeneinspannung, Forschungsbericht, Darmstadt 1983.

(8) Leicher, E.: Theoretische Untersuchungen zum Knickverhalten von Wänden ohne Zugfestigkeit, Dissertation, Darmstadt (in Arbeit).

Wall/Floor-Slab Interaction in Brickwork Structures

by

Professor A.W. Hendry

Abstract

The paper summarises the results of three series of tests on sections of brickwork structure carried out with a view to determining the characteristics of wall/floor slab joints in terms of fixity and resulting eccentricity of loading in a wall. Methods of calculating eccentricity and wall strength are discussed and it is shown that in practical situations eccentricities are likely to be small and capacity reduction factors close to unity.

1. Introduction

In most current codes of practice allowance for slenderness and eccentricity on the compressive strength of masonry walls is made by the application of a capacity reduction factor calculated on a semi-empirical basis. The assessment of the structural eccentricity required in this procedure is usually by a simple rule of thumb which neglects the variables known to influence the situation. In fact, both the eccentricity of loading and the wall strength depend on the interaction between the walls and the floor slabs. Eccentricity depends on moment transmission between the two elements and thus on their relative stiffnesses as well as on the fixity of the joint. Wall strength is influenced by the effect of the floor slabs in controlling the end rotations at failure. In current design procedures this is allowed for by assuming an effective height of wall less than the actual storey height in conjunction with capacity reduction factors determined on the basis of hinged end conditions for the wall.

The problem is clearly one of considerable complexity and it is important to be able to judge the adequacy of simplified design procedures in relation to the actual behaviour of brickwork structures. Towards this end a number of tests and associated theoretical studies have been carried out and these are summarised in this paper together with a discussion of the implications of the results for the design of brickwork structures.

* University of Edinburgh, Scotland.

2. Summary of Test Results on Joint Fixity and Eccentricity

2.1 Tests on Structures with 215mm (8.5/8 in) Thick Wall

Tests on two storey, single bay structures of the type shown in fig. 1 were carried out by Colville (1977) on a full scale structure and repeated by Awni (1980,1982) on a half-scale equivalent. The experimental variables were slab loading, joint precompression and loading sequence.

As will be seen from figs. 2 and 3, a high degree of fixity is attained in this form of wall/floor-slab joint at precompressions above 0.3-0.4 N/mm² (43-58 16/in²) but the fixity does not increase very much with precompressions in excess of this figure. It was found in these tests that the loading sequence had no effect on the joint behaviour.

A series of tests with similar objectives but with a different experimental arrangement has been reported by Stokle (1983)

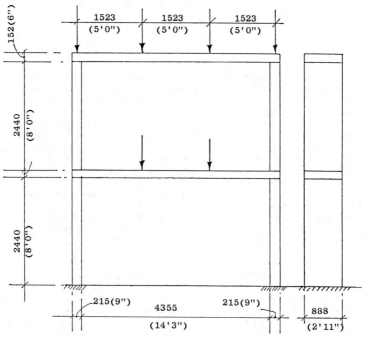

Fig. 1:   Test Structure   (Awni)

WALL/FLOOR-SLAB INTERACTION

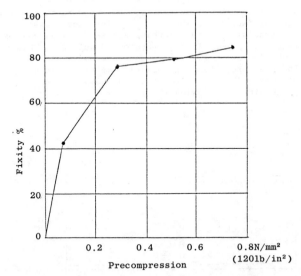

Fig. 2: Joint Fixity v precompression for 215mm wall structure.

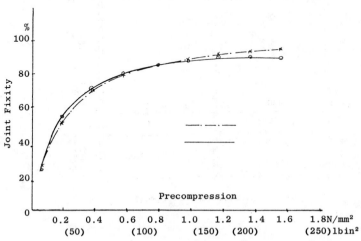

Fig.3: Joint fixity precompression for half scale structure with equivalent 215mm thick walls.

2.2 Tests in Structures with 103 mm (4in.) Thick Walls

Loading tests on the three-storey, two-bay structure shown in fig.4 were conducted by Awni (1980,1982). In these tests, attention was concentrated on joint C.2 and a jacking system was arranged so that precompression could be applied to the top of wall C. This permitted the evaluation of joint fixity at various precompressions, as shown in fig. 5. In this case, the joint fixities attained are quite low - of the order of 30%-even at relatively high precompressions. Similar results were obtained by Stokle (1983).

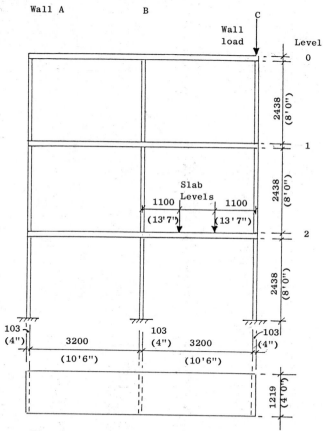

Fig. 4:   Test structure (Awni)

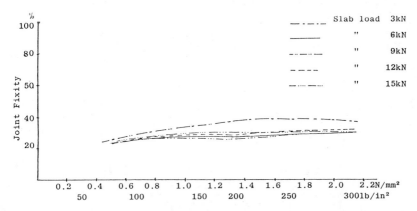

Fig. 5: Joint fixity v precompression for 103mm wall (Awni)

Tests were also carried out on a single bay structure with 103 mm (4in.) thick walls (fig. 6) by Chandrakeerthy (1983), the results of which can be seen in fig. 7. Allowing for the difference in the form of the test structures and the loading, these results are consistent with those obtained by Awni in that the fixity obtained between a floor slab and a 103 mm (4in.) thick wall is of the order of 30% at a precompression of 1 N/mm^2 (145 lb./in^2).

2.3 Tests on a Cavity Wall Structure

The behaviour of a cavity wall/floor-slab joint was examined by Chandrakeerthy (1983) in the structure shown in fig. 6. The result of these tests are summarised in fig. 7. which shows that at precompressions of up to about 0.3 N/mm^2(43lb/in^2) the joint was essentially rigid. At a slightly higher precompression, however, the fixity suddenly decreased and at higher precompressions the fixity was similar to that developed in the corresponding single leaf tests. It is clear that the sudden decrease must have resulted from cracking of the outer leaf which, immediately prior to cracking, would have been subjected to a high bending moment from the floor slab with only nominal compression from above.

STRUCTURAL MASONRY ANALYSIS

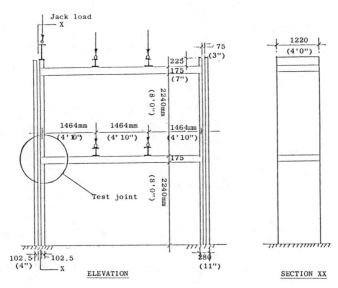

Fig. 6: Test Structure (Chandrakeerthy)

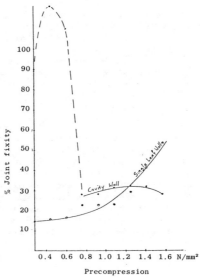

Fig.7: Joint fixity v precompression.

WALL/FLOOR-SLAB INTERACTION   93

3.0 The Calculation of Eccentricity

3.1 Eccentricity by Conventional Rule

Codes of practice for the design of masonry structures give simple rules for calculating eccentricity. Thus the British Code of Practice, BS.5628, Part 1 (1978) states that eccentricity should preferably be calculated but, at the discretion of the designer, it may be assumed that the load transmitted to a wall acts at one third of the depth of the bearing area from the loaded face of the wall. This of course has to be combined with loads from storeys above and the result may or may not be realistic. Wall/floor-slab interaction is taken into account in a general way by taking the effective height of a wall as three-quarters of the actual height.

3.2 Eccentricity from Partial Frame Analysis

Two approaches have been suggested for making a more realistic assessment of eccentricity. The first of these is to employ a partial frame analysis, as proposed by Haller (1969) and by Vahakallio and Makela (1975). Haller's formulae are shown in fig. 8 for the case of an outer wall. These methods are quite easily applied but are besed on the assumption of a rigid wall/floor-slab joint which could result in considerable over-estimation of eccentricity. Using the results of the tests described above, it is possible to introduce a modification to allow for partial fixity of the joint which is a function of the stiffness ratio of the wall and floor slab and the compressive stress on the joint. Thus fig. 9 shows experimentally determined values for the joint rigidity plotted against the ratio of slab to wall stiffness for various values of compression.

3.3 Wall Eccentricity from Floor Slab Interaction Formula

The second approach to the estimation of eccentricity follows that initiated by Sahlin (1971) and Risager (1969) and subsequently developed by Colville (1977) and Awni (1980). The method is based on consideration of the stressing of an equivalent column, assuming linear elastic material of zero tensile strength. Awni (1980) obtained the following formula for the eccentricity ratio:

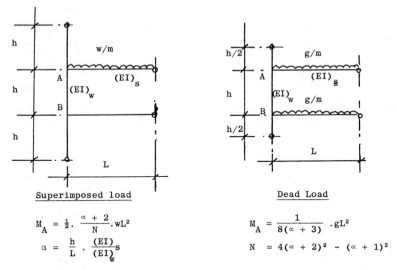

$$M_A = \tfrac{1}{2} \cdot \frac{\alpha + 2}{N} \cdot wL^2 \qquad M_A = \frac{1}{8(\alpha + 3)} \cdot gL^2$$

$$\alpha = \frac{h}{L} \cdot \frac{(EI)_s}{(EI)_w} \qquad N = 4(\alpha + 2)^2 - (\alpha + 1)^2$$

Fig. 8: Calculation of Eccentricity in Exterior Masonry Walls

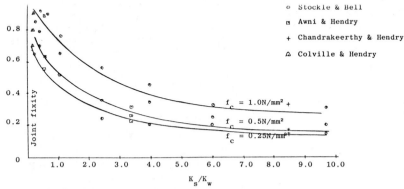

Fig. 9: Joint fixity v slab/wall stiffness ratio

$$\varepsilon = \frac{\bar{M}}{P_L t \bar{\phi}}$$

where $\bar{M} = wL^2/12$

$\bar{\phi} = [(1+\psi)(1+\bar{\beta})+\bar{K}/R]$

$\psi = P_u/P_L$

P_u and P_L are the loads above and below the joint under consideration

$\bar{\beta} = 2(EI)_s / BL$

β = Joint stiffness

$\bar{K} = 2(EI)_s H/(EI)_w L$

t = wall thickness

$(EI)_s/L$ and $(EI)_w/H$ are the stiffnesses of the slab and wall respectively.

Values of β for two wall thicknesses and a range of joint compressive stresses derived from experimental results are shown in Fig. 10.

The factor R depends on whether the wall is bent in single or double curvature and has the following values:

1. Walls bent in single curvature R = 1.85
2. Walls bent in double curvature

    (a) Uncracked (i.e. $\varepsilon < 1/6$) R = 2.15
    (b) Cracked   (i.e. $\varepsilon > 1/6$) R = 2.32

Comparisons between calculated and experimental values from Awni's tests, shown in figs. 11 and 12, indicate reasonably good agreement. There is also satisfactory agreement between the eccentricity ratios calculated in this way and by Haller's formulae, provided that the appropriate value of the joint rigidity is applied.

96   STRUCTURAL MASONRY ANALYSIS

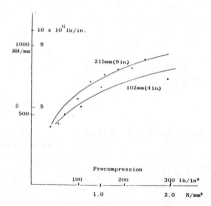

Fig. 10: Parameter β v joint precompression

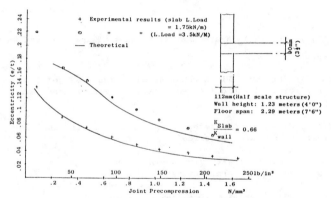

Fig.11: Experimental and theoretical eccentricities for half scale structure with equivalent 215mm (8½") thick walls.

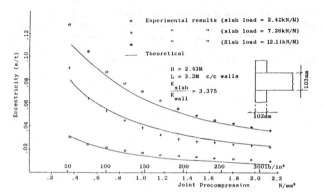

Fig.12 : Experimental and theoretical eccentricities for full scale structure with 102mm (4") thick walls

4. Calculation of Wall Strength

4.1 Capacity Reduction Factor Approach

In most codes of practice wall strength is calculated by multiplying the masonry compressive strength by a reduction factor which allows for second order effects arising from slenderness and eccentricity. The restraining effect of floor slabs is allowed for in a general way by calculating the slenderness ratio of the member on the basis of an "effective length". In some cases different reduction factors are used according to the eccentricities at the top and bottom of the wall.

Many theoretical solutions have been produced for this problem based on a variety of assumptions concerning material properties, initial irregularity of wall or load alignment and mathematical approximations. Most have involved solution of the differential equation for a brittle column and, more recently, computer simulation of the behaviour of the element has been undertaken (Sawko & Towler 1982). A few investigators have attempted to include the effect of wall/floor slab interaction but the majority are based on the assumption of hinged end conditions and rely on estimation of effective length to allow for end restraint.

The results of these analytical solutions for the case of a wall with an eccentric load at the top and zero eccentricity at the base are compared in Fig. 13. These are (a) a formal solution by Dr Groot and Van Riel (1967) (b) a solution including the effect of wall/floor slab interaction by Awni and Hendry (1981) and (c) a computer simulation by Sawko and Towler (1982). The computer solution modelled the same hinged - end system as assumed by De Groot and Van Riel and gives generally similar results. The solution including the effect of wall/floor slab interaction gives higher values of the capacity reduction factor at the smaller eccentricity ratio, as might be expected. However, if the slenderness ratio is based on an effective height equal to 3/4 of the actual height there is reasonable agreement between the various solutions.

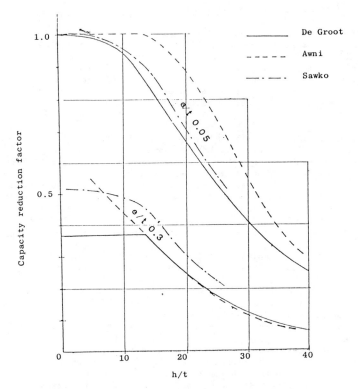

Fig. 13: Comparison between capacity reduction factors calculated by different methods.

4.2 Moment Magnifier Method

An alternative approach allowing for second order effects is the moment m gnifier method which is well known in North America. Its application to masonry elements has been discussed by Hatzinikolas et al (1982). In this method, an interaction curve defining the strength of the compression member subject to a combin
bending moment is derived. This may be done on the basis of a linear or non-linear stress-strain relationship. End conditions and slenderness are allowed for by applying a "moment magnifier" to the moment produced by the axial load P acting at eccentricity e. Thus:

WALL/FLOOR-SLAB INTERACTION

$$P(e + \Delta) = P \cdot e \cdot \frac{C_m}{1-P/P_{cr}}$$

where Δ is the maximum deflection of the element

$$C_m = 0.6 + 0.4\ e_1/e_2 > 0.4$$

e_1, e_2 are the eccentricities at top and base ($e_1 < e_2$)

$$P_{cr} = 8\pi^2\ \frac{EI_o}{h^2}\ (0.5-e/t)^3$$

in which I_o is the moment of inertia based on the gross cross section.

For a short wall the interaction diagram is defined by the following equation:

$$M = (P_o - P)\frac{t}{6} \quad \text{for } P/P_o > 0.5$$

$$M = (1 - \frac{4}{3}\frac{P}{P_o})\ \frac{P \cdot t}{2} \quad \text{for } P/P_o < 0.5$$

where
M = resisting moment
P = vertical load
P_o = compressive strength of the section

Points on the interaction diagram for a slender wall are determined by an interactive procedure, starting with values, say P_1 and M_1 for a short member. The critical load P_{cr} is then calculated and a first estimate of the axial load capacity of the slender member obtained by putting the left hand side of the moment magnifier equation equal to $P_1 \cdot e$ and $P = P_2$ on the left hand side. This procedure is repeated using P_2 and so on until convergence is obtained.

Experimental confirmation of the moment magnifier method appears to be somewhat inconclusive and comparison between wall strengths calculated by the reduction factor methods and the moment magnifier approach suggests that the latter gives considerably higher values. This is illustrated in Fig. 14 in the form of a load/moment interaction diagram. Some further investigation is necessary to reconcile the various methods.

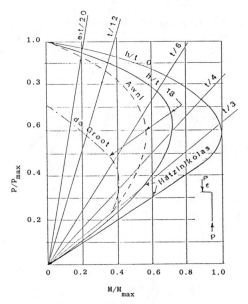

Fig. 14: Load/moment interaction curves by different methods.

5. Calculated Eccentricities and Reduction Factors in Practical Situations

As a means of demonstrating the practical signifance of eccentricity and related capacity reduction factors obtained by the methods of paragraphs 3 and 4 above, calculations have been carried out relating to the outer wall of the hypothetical building shown in Fig. 15.. A storey height of 2.55m (8ft 4in) has been assumed and two wall types are considered, namely a 280mm (11in.) cavity wall and a 215mm (8.5/8in) solid wall. The floor spans are taken as 3.5m (11ft 6in.) 4.5m (14ft 9in) with corresponding thicknesses of 150 (6in.) and 170mm (6.3/4 in) respectively. In calculating the eccentricities for the cavity wall case, it has been asumed on the basis of test results that the wall will behave as a single leaf wall when the joint precompression exceeds $0.3 N/mm^2$ ($43 lb/in^2$) thus only at the first level below the roof slab has the full stiffness of the wall been assumed. Capacity reduction factors obtained, by the method of Awni and Hendry (1983), allowing for wall floor slab interaction and from the British Code of Practice have been determined.

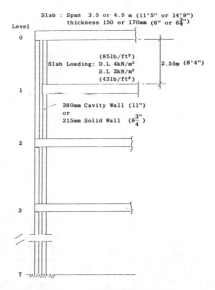

Fig.15: Dimensions and loading for outer wall of hypothetical building

The results are summarised in Table 1 from which the following points will be noted:
1. In general eccentricities are quite small but are appreciable in cavity walls at the first level below the roof, resulting from the relatively high stiffness of the wall at this level. At levels below this only one leaf will be effective at the joints with the result that the eccentricities are much reduced.

2. Because of the greater stiffness of the solid wall, the eccentricities at all levels are much higher than in the cavity wall, apart from level 1. The calculated values are consistent with experimental results on structures with walls of corresponding types.

3. In general, the calculated eccentricities are much larger than those indicated by the conventional rule, the difference being most pronounced when the wall stiffness is large.

4. The capacity reduction factors given by the two methods are effectively the same in nearly all cases. This is because the eccentricities given by the British Code method are negligible whilst the reduction factors even for appreciable eccentricities are close to unity in the alternative method.

Table 1

Eccentricities and Capacity Reduction Factors for Outer Wall of a Building

Level below Roof Slab	f_c N/mm²	Floor Slab : Span 3.5m									Floor Slab : Span 4.5m								
		280mm thick Cavity Wall				215mm thick Solid Wall				f_c N/mm²	280mm thick Cavity Wall				f_c N/mm²	215mm thick Solid Wall			
		e		B		e		B			e		B			e		B	
		Calc.	Code+	Calc.	Code+	Calc.	Code+	Calc.	Code+		Calc.	Code+	Calc.	Code+		Calc.	Code+	Calc.	Code+
1	0.2	0.13	0.063	0.97	0.88	0.17	0.06	0.96	0.95	0.12	0.22	0.067	0.72	0.88	0.31	0.20	0.066	0.85	0.88
2	0.5	0.075	0.038	1.0	0.9	0.16	0.038	0.98	0.98	0.24	0.088	0.041	1.0	0.9	0.68	0.14	0.04	1.0	0.9
3	0.75	0.057	0.028	1.0	0.9	0.14	0.027	1.0	0.98	0.36	0.069	0.030	1.0	0.9	0.93	0.12	0.03	1.0	0.9
4	1.0	0.046	0.023	1.0	0.9	0.11	0.022	1.0	0.98	0.49	0.055	0.023	1.0	0.9	1.24	0.094	0.023	1.0	0.9
5	1.25	0.038	0.018	1.0	0.9	0.094	0.018	1.0	0.98	0.61	0.047	0.020	1.0	0.9	1.54	0.078	0.02	1.0	0.9
6	1.5	0.034	0.015	1.0	0.9	0.079	0.015	1.0	0.98	0.73	0.041	0.017	1.0	0.9	1.85	0.067	0.016	1.0	0.9
7	1.75	0.025	0.013	1.0	0.9	0.060	0.013	1.0	0.98	0.85	0.031	0.015	1.0	0.9	2.16	0.069	0.015	1.0	0.9

f_c = precompression on joint e = eccentricity ratio B = slenderness and capacity reduction factor

+ = British Standard 5628 Part 1 (1978)

6. Conclusion

The rigidity of wall/floor slab joints is influenced by the relative stiffness of the wall and floor slab and by the compression from floors above. It has been confirmed by extensive tests on sections of full scale brickwork structure that the structural eccentricity resulting from wall/floor slab interaction can be calculated with reasonable accuracy by comparatively simple methods.

The eccentricity at the top of a section of wall in a multi-storey brickwork building is in general, quite small but may be appreciable in a flank wall at the upper floor levels. At these levels, however, the compressive stress in the brickwork will be small.

The strength of an eccentrically loaded wall can be estimated by the use of capacity reduction factors calculated by the solution of the differential eqation for a brittle column or by computer simulation. If the solution is based on the assumption of hinged end conditions the assumption of an effective height of 0.75 of the clear height between floor slabs will be reasonably accurate. In most practical cases the reduction factor will be close to unity, at least when the compressive stress in the brickwork is significant.

Acknowledgements

The experimental work referred to in this paper was undertaken with funds contributed over a number of years by the British Ceramic Research Association and the Brick Development Association.

REFERENCES

AWNI, A.A. — "The Compressive Strength of Brick Masonry Walls with Reference to Wall/Floor Slab Interaction", Ph.D Thesis. University of Edinburgh, 1980.

AWNI, A.A. & HENDRY, A.W. — "The Strength of Masonry Walls Compressed between Floor Slabs", Int.J. Masonry Construction, Vol.1 No.3, 1981, pp.110-118.

AWNI, A.A. & HENDRY, A.W. — "Joint Fixity Measurements on Load Bearing Masonry Structures", Proc. B. Ceram. Soc. No.30, 1982, pp. 149-159.

AWNI, A.A & HENDRY, A.W. — "A Simplified Method for Eccentricity Calculation", Proc. Vth Int. Brick Masonry Conf., Ed. J.A.Wintz and A.H.Yorkdale, Washington, 1982, pp. 528-534.

B.S.I. — "Code of Practice for Structural Use of Masonry, BS 5628 Part 1, Unreinforced Masonry, British Standards Institution, London, 1978.

CHANDRAKEERTHY, S.R. DE, & HENDRY, A.W. — "Behaviour of Wall-Floor Slab Joints Leaf and Cavity Walls", Int. J. of Masonry Construction, Vol.3 No.1 1983, pp 10-13.

COLVILLE, J. — "Analysis and Design of Brick Walls", University of Edinburgh, Department of Civil Engineering and Building Science, June 1977.

COLVILLE, J & HENDRY, A.W. — "Tests of a Load Bearing Masonry Structure", Proc. B. Ceram. Soc: No.27. 1978, pp 77-84.

De GROOT A.K. & Van RIEL, A.C. — "De Stabiliteit van Kolommen en Wanden van Ongewapend Beton",Heron, Jaargang 15, No. 3/4, Delft, 1967, pp.65-107.

HALLER, P — "Load Capacity of Brick Masonry", Designing Engineering and Constructing with Masonry Products, ed. F.B. Johnson, Gulf Pub. Co., Houston, 1969, pp. 129-149.

HATZINCKOLAS, M.A. LONGWORTH, J. & WARWARUK, J. — "Design of Slender Brick Elements Based on Stability Criteria", Proc. B.Ceram. Soc. No.30, 1982, pp. 236-245.

RISAGER, S. — "Structural Behaviour of Linear Elastic Walls having no Tensile Strenth" , Designing, Engineering and Constructing with Masonry Products, ed. F.B. Johnson, Gulf Pub. Co., Houston, 1969, pp 257-265.

SAHLIN, S. — "Structural Masonry", Prentice Hall, 1971, pp.91-116

SAWKO, F. &   "Numerical Analysis of Walls and Piers with
TOWLER K.   Different End Conditions", Proc. 6th Int. Brick
                 Masonry Conference, Rome, 1982, pp. 412-421.

STOKLE, J.D.   "Structural Interaction between Reinforced Concrete
                 Floor Slabs and Plain Masonry Walls", M.Sc. Thesis,
                 University of Manchester Institute of Science and
                 Technology, 1983.

VAHAKALLIO, P. & "Method for Calculating Restraining Moments in
MAKELA, K.   Unreinforced Masonry", Proc. B.Ceram.Soc., No.24,
                 1975, 161-175.

SHEAR STRESSES IN COMPOSITE MASONRY WALLS

Subhash C. Anand[1], M.ASCE

ABSTRACT

The behavior of composite masonry walls subjected to inplane loads is a subject that has received little attention during the last few years. The small amount of experimental and analytical research performed on composite masonry has not offered insight into the understanding of interlaminar shearing stresses in the collar joint that are created by the interaction between the two wythes of masonry. Determination of these stresses is quite important for the safe and economic design of composite masonry structures.

As a part of the research at Clemson University, that has been supported by NSF, a two-dimensional composite finite element model has been developed that is capable of predicting the out-of-plane interlaminar shearing stresses between the brick and block wythes without using a three-dimensional finite element model. This composite element is utilized in two-dimensional analyses of a composite masonry wall with window opening subjected to floor loads on the block wythe, that could not have been analyzed using a standard plane-strain model. The proposed model is also employed in the analysis of a laboratory specimen, results of which are compared with those of the plane-strain model. It is found that the proposed composite element model (longitudinal model) yields shear stresses in the collar joint near the point of load application that are much higher than those predicted by the plane-strain model. An appropriate shear stress distribution in the collar joint is also obtained from a closed-form elasticity solution which qualitatively agrees with the plane-strain finite element results. It is concluded, therefore, that the proposed composite element model is quite useful in estimating shear stresses in composite walls with openings or with discontinuous loads when plane-strain finite element model cannot be employed. However, it yields conservative values for transverse shear stresses in the collar joint. Transverse plane-strain finite element model or closed-form elasticity solution may be used in cases where these are applicable. The collar joint shear stresses have been obtained for vertically applied loads only. Further investigations must be conducted with horizontally applied in-plane loads on composite walls. Initiation and propagation of cracks in the collar joint and the criteria for separation of the two wythes must be further studied.

---

[1]Professor of Civil Engineering, Clemson University, Clemson, SC, 29634-0911, USA

INTRODUCTION

A composite wall consists of a layer (wythe) of brick and, closely adjacent, a layer (wythe) of concrete block. If the cavity (collar joint) between the two wythes remains hollow, as shown in Fig. 1(a), then the structural properties of each wythe are independent of each other. When the collar joint is filled with grout and steel ties, as shown in Fig. 1(b), the composite wall becomes a complex structural assemblage with mechanical properties that are a function of the wall's constituents, i.e., concrete block, brick, and grout.

In many practical applications, composite masonry walls carry substantial loads. Figure 2 shows a floor slab which has been placed upon the concrete block wythe of a composite wall. The slab can apply horizontal and vertical loads to the block wythe; these loads are then distributed through the collar joint to the brick wythe. This transfer occurs through the collar joint primarily as a shear stress (1-4). The presence of large shear stresses in the collar joint could cause failure of the wall. If the shear stresses become too large at the brick-mortar interface or at the block-mortar interface, delamination may occur. If this happens, the load will no longer be transferred to the brick wythe and the wall may be underdesigned. The primary objective of this research, therefore, is to determine the magnitude and distribution of these shear stresses.

Previous and Current Research

An extensive and indepth review of all previous and current research dealing with composite masonry walls can be found in Refs. (1-4). Only a brief description of the most recent research is presented here, which may be divided into two basic categories; namely, analytical (computational) and experimental.

The experimental research has focused its attention either on the compressive strength of prisms where both wythes are loaded (19, 20), or on the shear as well as compressive strengths of both the wythes and the collar joint. In the latter case, most of the tests have been performed on wallets which are subjected to inplane loads that produce normal stresses in the wythes and shear stresses in the collar joint (8, 15, 16, 18, 22). The research at Iowa State University (18), predicts certain combinations of axial and lateral loads at failure of either the collar joint or one of the wythes. On the other hand, and average shear stress at failure of the collar joint is predicted by the tests conducted at Penn State (22) and Clemson University (8, 15, 16), where variations in the strength values due to different materials used in the construction are demonstrated. It should be noted that it has as yet not been possible to determine the variation of shear stress in the collar joint experimentally.

Most of the analytical research on the behavior of composite masonry walls subjected to inplane loads has been conducted at Clemson University (1-7, 9-14) by this author and his graduate students. In

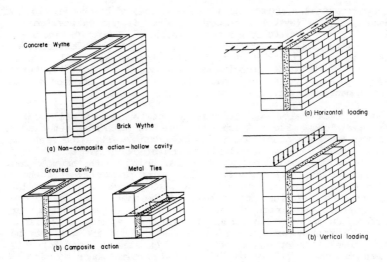

Fig. 1 Composite and Non-Composite Wall

Fig. 2 Loads on a Composite Wall

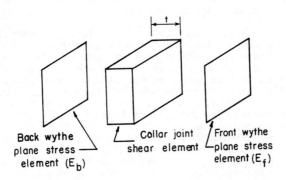

Fig. 3 Components of a Composite Masonry Finite Element

this research, the investigators have developed a 2-dimensional finite element model that is capable of predicting the normal stresses in the wythes in addition to the shear stresses in the collar joint (1-4, 6, 14). The same model has also been utilized to estimate stress and strain variations in composite walls due to creep (5, 7, 9, 11, 13). In addition, the 2-D model has been extended to incorporate shear cracking and failure of the collar joint (6, 10, 12). The only other analytical work that suggests a methodology to predict shear stresses in the collar joint (when both wythes of a composite wall are loaded) is due to Grimm and Fowler (17) that uses a simple mechanics of materials approach.

Although the development of the proposed 2-D model has been presented elsewhere previously (1-4, 6,14), some important features of the same are given in the next section for completeness. The proposed 2-D model is utilized subsequently to compute shear stresses in a composite wall with a window opening which could not be analyzed using the standard two-dimensional plane strain model. It is shown that the model is capable of predicting the normal and shear stresses in the wythes as well as in the collar joint.

In order to verify the accuracy of the shear and normal stresses in a composite wall predicted by this model, a long composite wall, without any wall opening and subjected to a continuous uniform in-plane load, is analyzed using the proposed model as well as a standard plane-strain finite element model. It is apparant from the results that, although the proposed 2-D model is extremely useful in the analysis of discontinuous walls or walls with openings, it predicts higher shear stresses in the collar joint near the point of application of the load. The shear stresses in the collar joint are also predicted using the closed-form elasticity solution of an half space which agree qualitatively with those predicted by the plane-strain finite element model. It is concluded, therefore, in this paper that the longitudinal shear stresses in the collar joints of composite walls subjected to discontinuous in-plane loads can be estimated fairly accurately using the proposed 2-D model whereas the simple plane strain model of the wall cross-section can be utilized for estimating collar joint shear stresses across the wythes in long composite walls that are subjected to continuous distributed loads on one wythe. Although some preliminary research on crack initiation and propagation in the collar joint due to shear stresses has also been conducted by this author and his students (6, 10, 12), it will not be reported in this paper.

DESCRIPTION OF THE TWO 2-D MODELS

Transverse Model

A composite wall subjected to in-plane loads can be modeled as a plane strain case by considering only its cross-section if the wall is long, and the loading is uniform and continuous along the length of the wall. This type of model is termed as the Transverse Model in this paper.

Longitudinal Model

The other type of 2-D model that is proposed is named as the Longitudinal Model and is developed to overcome the shortcomings of the standard 2-D and 3-D models. Cross-sectional 2-D analyses are easily performed but cannot be used efficiently to model the discontinuities found in real walls such as doorways, windows, and partial loadings. A 3-D model can handle these discontinuities but the analysis requires large amounts of computer time and the results are difficult to evaluate due to the complexity of the model. The proposed longitudinal model incorporates the positive aspects of 2-D and 3-D models without the associated restrictions. Some description of the longitudinal model is given as follows:

In the development of the longitudinal model, a new "composite" element is created. In this new element, the front and back wythes are each modelled as plane stress elements. These two elements are joined together by a collar joint shear element. These three elements are shown in Fig. 3, a combination of which forms the composite element which has eight nodes — four for each wythe -- as shown in Fig. 4. As each node has two degrees of in-plane displacement freedom, there are in total sixteen degrees of freedom for the element. The shear stresses that act in the collar joint shear element are shown in Fig. 5.

The stiffness matrix of the proposed composite element is formed by combining the stiffness matrices of the two wythe elements with the collar joint stiffness matrix. Detailed expressions for these matrices are derived in the following sections. It should be noted that the following assumptions have been made in this development. 1) All materials are considered as elastic, homogeneous, and isotropic; 2) Displacements are assumed to vary linearly between nodes in an element; 3) Out-of-plane bending effects in the wall are ignored; and 4) The collar joint as well as the two wythes are assumed to be unreinforced.

Stiffness Matrix of a Wythe Element

Determination of stresses and displacements in wythes due to in-plane loads can be accomplished by the standard plane stress finite element analysis. The governing matrix equation relating forces and displacements in an element is given by

$$\{P\} = [k]\{U\} \qquad (1)$$

where $\{P\}$ is a column matrix of force-components in the x and y directions at the nodes of an element, $[k]$ is the in-plane stiffness matrix for an element, and $\{U\}$ represents the corresponding displacement components at the nodes. A more detailed development of the in-plane stiffness matrix may be found in standard finite element texts. Quadrilateral elements, each consisting of four triangular elements, are utilized in subdividing the wythe faces into a finite

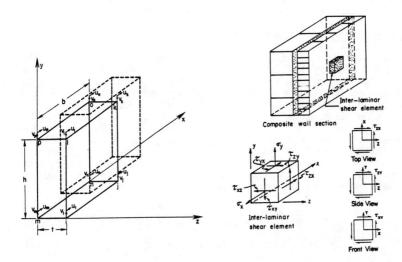

Fig. 4 Nodal Degrees of Freedom

Fig. 5 Interlaminar Shear Stresses

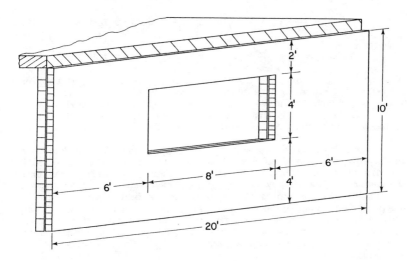

Fig. 6 Twenty Foot Section of a Long Wall

element mesh. As each node has two degrees of freedom, Eq. 1 yields an 8 x 8 stiffness matrix, [k], for each element.

Stiffness Matrix of a Collar Joint Element

The collar joint element stiffness provides the interaction between the two plane stress elements representing the masonry wythes. The three parts of a composite element are shown in Fig. 3 with the wythe element stiffnesses for the front wythe $[k_f]$ and the back wythe $[k_b]$ having been developed in the previous section. Shear deformation of the collar joint element is composed of displacements in the x and y directions only. Since these displacements are at the nodes of the plane-stress grids and vary linearly in the plane of each grid, the stiffness properties of the collar joint elements are also based on a linear displacement field in the x - y plane. Thus, the strain-displacement relations in the collar joint element may be written in terms of the nodal displacements as

$$\varepsilon_x = \frac{\partial u}{\partial x} = (u_j+u_k+u_n+u_o-u_i-u_l-u_m-u_p)/4b \qquad (2)$$

$$\varepsilon_y = \frac{\partial u}{\partial y} = (v_k+v_l+v_o+v_p-v_i-v_j-v_m-v_n)/4h \qquad (3)$$

$$\gamma_{xy} = \frac{\partial u}{\partial y} + \frac{\partial v}{\partial x} = (u_k+u_l+u_o+u_p-u_i-u_j-u_m-u_n)/4h +$$

$$(v_j+v_k+v_n+v_o-v_i-v_l-v_m-v_p)/4b \qquad (4)$$

$$\gamma_{zx} = \frac{\partial u}{\partial z} + \frac{\partial w^0}{\partial x} = (u_i+u_j+u_k+u_l-u_m-u_n-u_o-u_p)/4t \qquad (5)$$

$$\gamma_{zy} = \frac{\partial v}{\partial z} + \frac{\partial w^0}{\partial y} = (v_i+v_j+v_k+v_l-v_m-v_n-v_o-v_p)/4t \qquad (6)$$

in which u and v are the displacements in the x and y direction, respectively, subscripts refer to the specific element nodes, and t, h, and b, are the thickness, height, and length, respectively, of the collar joint element as shown in Fig. 4. Note that displacement, w, in the z direction is not allowed in this model.

The shearing strains γ_{zx} and γ_{zy} in these equations are defined as the average relative displacement between the two wythes divided by the distance between the two wythes. Thus, the medium resisting shear across the two wythes may be considered as a shear segment connecting the centroids of the two elements facing each other. The shear strain, γ_{xy}, on the other hand, is defined by the average relative y-displacements of the nodes on the two x-faces of the composite element divided by the length of the x-element in the x-direction. The two normal strains, ε_x and ε_y, can be defined similarly.

For a general quadrilateral (i.e., when the shape of the wythe element is other than rectangular), t remains constant; however, b and h are defined as

$$b = (x_j + x_k - x_l - x_i)/2$$
$$h = (y_k + y_l - y_i - y_j)/2 \qquad (7)$$

in which subscripted x and y are the global nodal point coordinates of the corresponding nodes. Thus, b and h represent the average length and height of an element, respectively.

Equation 2-6 may be rewritten in the matrix form as

$$\{\varepsilon\} = [B]\{U\} \qquad (8)$$

where

$$\{\varepsilon\} = [\varepsilon_x \varepsilon_y \gamma_{xy} \gamma_{zx} \gamma_{zy}]^T \qquad (9)$$

$$[B] = \begin{bmatrix} \frac{-1}{b} & 0 & \frac{1}{b} & 0 & \frac{1}{b} & 0 & \frac{-1}{b} & 0 & \frac{-1}{b} & 0 & \frac{1}{b} & 0 & \frac{1}{b} & 0 & \frac{-1}{b} & 0 \\ 0 & \frac{-1}{h} & 0 & \frac{-1}{h} & 0 & \frac{1}{h} & 0 & \frac{1}{h} & 0 & \frac{-1}{h} & 0 & \frac{-1}{h} & 0 & \frac{1}{h} & 0 & \frac{1}{h} \\ \frac{-1}{h} & \frac{-1}{b} & \frac{-1}{h} & \frac{1}{b} & \frac{1}{h} & \frac{1}{b} & \frac{1}{h} & \frac{-1}{b} & \frac{-1}{h} & \frac{-1}{b} & \frac{-1}{h} & \frac{1}{b} & \frac{1}{h} & \frac{1}{b} & \frac{1}{h} & \frac{-1}{b} \\ \frac{1}{t} & 0 & \frac{1}{t} & 0 & \frac{1}{t} & 0 & \frac{1}{t} & 0 & \frac{-1}{t} & 0 & \frac{-1}{t} & 0 & \frac{-1}{t} & 0 & \frac{-1}{t} & 0 \\ 0 & \frac{1}{t} & 0 & \frac{1}{t} & 0 & \frac{1}{t} & 0 & \frac{1}{t} & 0 & \frac{-1}{t} & 0 & \frac{-1}{t} & 0 & \frac{-1}{t} & 0 & \frac{-1}{t} \end{bmatrix}$$

(10)

and

$$\{U\} = [u_i\ v_i\ u_j\ v_j\ ...u_p\ v_p]^T \qquad (11)$$

The stress-strain relations, in this case, may be given by

$$\begin{Bmatrix} \sigma_x \\ \sigma_y \\ \tau_{xy} \\ \tau_{zx} \\ \tau_{zy} \end{Bmatrix} = \frac{E}{1-\nu^2} \begin{bmatrix} 1 & \nu & 0 & 0 & 0 \\ \nu & 1 & 0 & 0 & 0 \\ 0 & 0 & \frac{1-\nu}{2} & 0 & 0 \\ 0 & 0 & 0 & \frac{1-\nu}{2} & 0 \\ 0 & 0 & 0 & 0 & \frac{1-\nu}{2} \end{bmatrix} \begin{Bmatrix} \varepsilon_x \\ \varepsilon_y \\ \gamma_{xy} \\ \gamma_{zx} \\ \gamma_{zy} \end{Bmatrix} \qquad (12)$$

which yields the material property matrix, [D], as

$$[D] = \frac{E}{1-\nu^2} \begin{bmatrix} 1 & \nu & 0 & 0 & 0 \\ \nu & 1 & 0 & 0 & 0 \\ 0 & 0 & \frac{1-\nu}{2} & 0 & 0 \\ 0 & 0 & 0 & \frac{1-\nu}{2} & 0 \\ 0 & 0 & 0 & 0 & \frac{1-\nu}{2} \end{bmatrix} \qquad (13)$$

As in the case of the in-plane stiffness matrix for an element, the force-displacement relations for the collar joint shear element may be given by Eq. 1, in which the element stiffness matrix [k] is defined as

$$[k] = \int_{vol} [B]^T [D][B] dV. \qquad (14)$$

Carrying out the matrix multiplication above leads to the collar joint shearing element stiffness matrix, $[k_{sh}]$.

Stiffness Matrix of a Composite Element

The superposition of the two 8 x 8 wythe element stiffness matrices and the 16 x 16 collar joint shear element stiffness matrix results in a 16 x 16 composite element stiffness matrix which is given by

$$\begin{bmatrix} [k_f] & [0] \\ 8 \times 8 & 8 \times 8 \\ [0] & [k_b] \\ 8 \times 8 & 8 \times 8 \end{bmatrix} + \begin{matrix} [k_{sh}] \\ 16 \times 16 \end{matrix} \qquad (15)$$

in which $[k_f]$ and $[k_b]$ are the plane stress stiffness matrices of the front and back wythes, respectively; and $[k_{sh}]$ is the stiffness matrix of the collar joint.

Calculation of Displacements, Stresses and Strains

Using the stiffness matrix of a composite element given in Eq. 15, the stiffness matrix for a finite element model of the total structure can be assembled by the standard methods leading to the equilibrium equations which are solved for the nodal point displacements. Normal and shearing strains in the wythe elements are obtained using the standard strain-displacement relations of 2-D quadrilateral elements, whereas the corresponding strains in the collar joint elements are calculated using Eqs. 2-6. Similarly, the in-plane stresses in the wythe elements are calculated from the in-plane strains by using the standard plane-stress stress-strain relations. The normal and shearing stresses in the collar joint, on the other hand, are calculated from the corresponding strains using Eq. 13.

COMPOSITE MASONRY WALLS 115

ANALYSIS OF A COMPOSITE WALL WITH WINDOW OPENING

To demonstrate the capability of the proposed model, a realistic example problem is solved to determine the stresses in the collar joint. A 20 foot section of a long composite wall is arbitrarily chosen and is shown in Fig. 6. A concrete floor slab rests on the 8 in thick concrete block wythe which is connected to the 4 in thick brick wythe by a 3/8 in collar joint. The wall contains an opening for a 4 foot by 8 foot window.

The finite element mesh, boundary conditions, and loading are shown in Fig. 7. Due to symmetry, only half the wall is modelled and vertical rollers are placed at the center line. The vertical rollers on the left face approximate the plane strain conditions found in a long wall. A typical load of 2 kips per foot is applied at the nodes of the plane stress elements of the block wythe. The mesh is refined near the upper left hand corner of the window where large stress concentrations are expected. It should be noted that the section of the wall above the window is similar to a deep, short, fixed end beam. The material properties used represent some typical values and are given in Table I.

Table I. Typical Material Properties

Masonry Type	E, ksi	ν
4 Inch Brick	2,000	0.25
8 Inch Hollow Block	1,000	0.25
Grout	1,800	0.20

Results and Discussion

Figure 8 shows the different sections of the wall where stress distributions are plotted. The collar joint stresses σ_x, σ_y, τ_{xy}, and τ_{zy} are shown in Figs. 9-12.

The collar joint normal stress in the longitudinal direction is shown in Fig. 9. At section A-A, the stress distribution is linear as would be expected from classical beam theory. The distribution at A-A is also shifted slightly into the compression region due to the horizontal compressive stresses created by the rollers on both sides of the wall. The distribution at B-B is also shifted to the left and has become nonlinear due to its proximity to the corner of the window. At sections C-C and D-D, the horizontal normal stresses are tensile at the top of the wall and become compressive near the upper left hand corner of the window. This reversal in the sign of the stress may be attributed to the fixed end effects of the wall above the window opening near sections C-C and D-D.

If one considers the 8 foot span above the window as a 2 foot deep beam fixed at ends made of a homogeneous material of 12-3/8 in width, then the normal stress σ_x can be computed from the beam theory. The maximum normal stress at the midspan (section A-A) is calculated as $\pm$ 7.75 ksf, whereas the corresponding values at the top and bottom at

116 STRUCTURAL MASONRY ANALYSIS

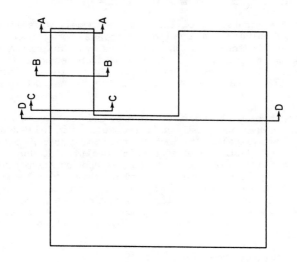

Fig. 8 Sections of Interest in the 20 Foot Wall

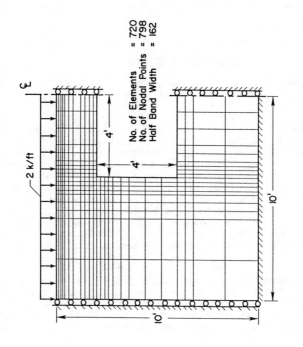

Fig. 7 Mesh, Loads, and Boundary Conditions for the 20 Foot Section of Wall

COMPOSITE MASONRY WALLS

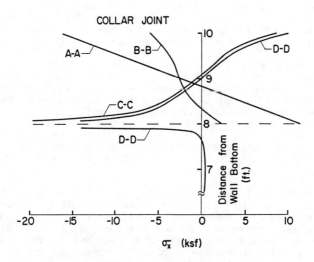

Fig. 9 Horizontal Normal Stress, σ_x, in the Collar Joint

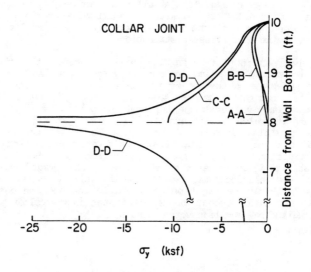

Fig. 10 Vertical Normal Stress, σ_y, in the Collar Joint

section D-D (assuming end fixity) are ±15.50 ksf. These values are reasonably close to those obtained using the finite element model, the latter being more accurate as the beam solution does not include the effects due to other portions of the wall.

The vertical stress in the collar joint is shown in Fig. 10. At sections A-A and B-B, σ_y is zero at the top and bottom of the beam, satisfying equilibrium at the surface. The stress in section C-C, on the other hand, is zero at the top and increases towards the corner of the window. If the model were more refined at the point where section C-C and the upper edge of the window meet, this stress would go to zero as it does in sections A-A and B-B. The stress in section D-D begins at zero at the top and approaches a peak at the corner of the window due to the stress concentration effects of the window opening.

The longitudinal shear stress, τ_{xy}, in the collar joint is shown in Fig. 11. At sections A-A and B-B, the stress distribution has the classic parabolic shape predicted by the beam theory. The shear stress distributions at sections C-C and D-D are zero at the top and climb to a peak at the corner of the window. It should be pointed out that it is not possible to predict the shear stresses at sections C-C and D-D using the beam theory.

The transverse shear stress, τ_{zy}, which is primarily responsible for transferring the vertical load from the block wythe to the brick wythe, is shown in Fig. 12. At all four sections, τ_{zy} reaches a peak at the top and drops off quickly to a small value at the 9 foot level. At the upper edge of the window opening, only a small deviation is seen among the four stress distributions.

From the results plotted in Figs. 9-12, it appears that the proposed model is capable of estimating correctly the normal and longitudinal shear stresses in the wythes as well as the collar joint. Figure 12 indicates that the transverse shear stress, τ_{yz}, in the collar joint has a maximum value at the top near the point of application of the load and its value remains essentially the same at all the sections. It is possible, therefore, that the presence of the wall openings could be ignored in the analysis as far as an estimate of the transverse shear stresses, τ_{xz}, τ_{yz} is concerned. However, a closer look at Fig. 12 indicates that the shear stress magnitude near the top of the wall decreases rapidly and the validity about its accuracy needs further investigation. The following section describes a comparison of the shear stress estimates in the collar joint using the longitudinal and transverse models.

COMPOSITE MASONRY WALLS

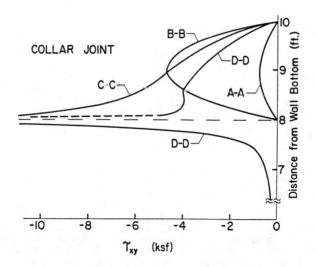

Fig. 11 Longitudinal Shear Stress, τ_{xy}, in the Collar Joint

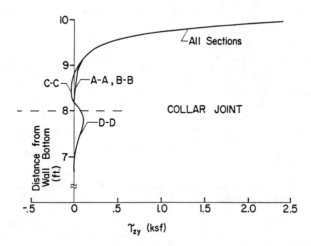

Fig. 12 Transverse Shear Stress, τ_{zy}, in the Collar Joint

COMPARISON OF RESULTS WITH THE TWO MODELS

Analysis of a Test Specimen

In order to assess the performance of the longitudinal model relative to the standard plane-strain transverse model, solutions for a laboratory specimen of the composite wall were obtained using the two models. Fig. 13 shows the plane strain element mesh of the transverse model subjected to a uniform load on the block wythe and Fig. 14 shows the longitudinal model with the boundary conditions that force it to a plane strain condition. Both models have the same vertical mesh refinement and load intensity. Material properties used in the analyses have been given previously in Table I.

The shear stress distributions found with each model are shown in Fig. 15, from which it is obvious that there is a large discrepancy in the results obtained with the two models. In the transverse model, the shear stress starts at the top of the collar joint with a zero value, as equilibrium requires the stress to be zero at a free surface. The shear stress then quickly increases to its maximum value and gradually drops to zero towards the bottom. The peak in the shear stress distribution is caused by the load transfer from the block wythe to the brick wythe through the collar joint. Fig. 16(a) shows the probable path of the load transfer. Most of the load remains in the block wythe while some load transfers into and through the collar joint as shear stress. As the load transfer occurs fastest near the point of load application, the shear stress in the collar joint goes from zero at the top to its peak value in a short distance. The shear stress then tapers off as the load is distributed into the two wythes and collar joint. It is important to note that this model allows displacements in- and out-of-plane of the wall.

In the derivation of the composite element stiffness in the longitudinal model, the shear stresses, τ_{zx} and τ_{zy}, in the collar joint are based strictly on the relative movements of the brick and block wythes in the vertical plane without any regard to the out-of-plane displacements. This constraint in the development of the longitudinal model can cause large shear stresses in the collar joint at the top of the wall where the true shear stress from equilibrium considerations must be zero. Consequently, the shear stress in Fig. 15 for the longitudinal model of the specimen is not zero at the top.

Fig. 16(b) shows a cross-sectional view of the longitudinal model in which the thicknesses of the brick and block wythes have been collapsed to emphasize the fact that only one element for the whole thickness has been used to model each wythe. The distributed load on the concrete block wythe in the transverse model is replaced by a point load positioned on the node of the element representing the concrete block wythe. The load may then travel two paths: into the block wythe elements or through the collar joint into the brick wythe. The stiffness of the block wythe is essentially that of an axial

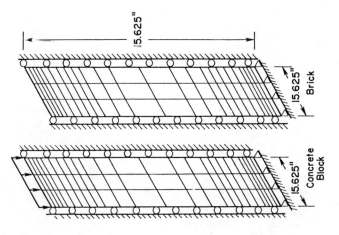

Fig. 14 Longitudinal Model of Test Specimen

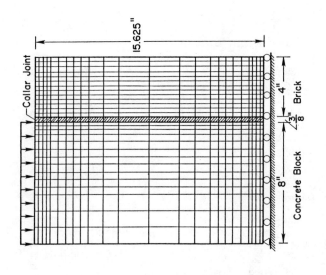

Fig. 13 Transverse Model of Test Specimen

122   STRUCTURAL MASONRY ANALYSIS

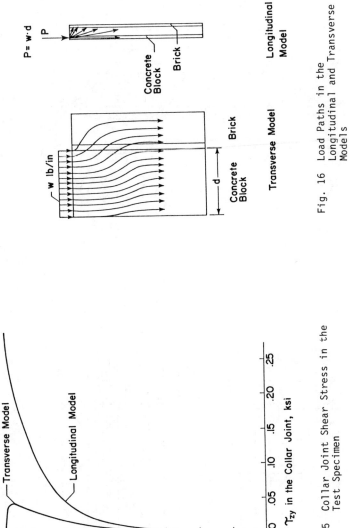

Fig. 15 Collar Joint Shear Stress in the Test Specimen

Fig. 16 Load Paths in the Longitudinal and Transverse Models

spring, since out-of-plane displacements are not allowed and displacements in the longitudinal direction have been restrained by the rollers shown in Fig. 14. The stiffnesses of the collar joint and brick wythe are also one dimensional for the same reason. Thus, the concrete block wythe, the collar joint, and the brick wythe are all overly stiff due to the disregard of the out-of-plane displacements, and a faster load transfer would be expected. It should be emphasized that the application of the uniformly distributed load of the block wythe as an equivalent concentrated load in the longitudinal model has a substantial influence on the magnitude of the maximum shear stress in the collar joint.

Due to the reasons given above, larger transverse shear stresses in the collar joint are predicted in the longitudinal model than in the transverse model. Although the results of the transverse model can be accepted without much doubt, some approximate verification of the shear stresses computed from this model can be achieved from extrapolation of an elasticity solution which is carried out in the next section.

Elasticity Solution

Although no closed-form elasticity solutions are available that can analyze composite walls loaded only on the block wythe, solution of a semi-infinite plate subjected to a uniform load may be used to estimate the accuracy of the finite element solution of the composite wall. A semi-infinite plane subjected to a uniform load, p, of width 2a is shown in Fig. 17. This is a stress boundary-value problem and its closed-form solution for stresses is given by (21)

$$\sigma_z = -\frac{p}{\pi} \{ \arctan(\frac{y}{z-a}) + \frac{(z-a)y}{(z-a)^2 + y^2}$$
$$- \arctan(\frac{y}{z+a}) - \frac{(z+a)y}{(z+a)^2 + y^2} \} \quad (16)$$

$$\sigma_y = -\frac{p}{\pi} \{ \arctan(\frac{y}{z-a}) - \frac{(z-a)y}{(z-a)^2 + y^2}$$
$$- \arctan(\frac{y}{z+a}) + \frac{(z+a)y}{(z+a)^2 + y^2} \} \quad (17)$$

$$\tau_{zy} = -\frac{p}{\pi} \{ \frac{y^2}{(z-a)^2 + y^2} - \frac{y^2}{(z+a)^2 + y^2} \} \quad (18)$$

The boundary condition on the shear stress is that $\tau_{yz}=0$ along $y=0$, which is satisfied by the above solution. However, if the points ($z=\pm a$, $y=0$) are approached along $z=\pm a$, then $\tau_{zy}(z=\pm a, y=0)=\mp p/\pi$. Therefore, $\tau_{zy} \neq \tau_{yz}$ at the points ($z=\pm a$, $y=0$) due to the stress discontinuity.

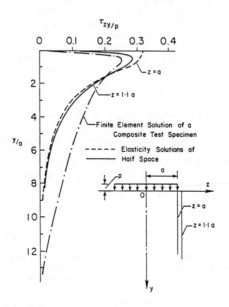

Fig. 17 Comparison of Half-Space Solution and Finite Element Solution

The normalized values of τ_{zy} along the y direction computed from Eq. 18 is plotted in Fig. 17 for fixed values of distances z=a and z=1.1a. It is seen from Eqs. 16-18 that the stresses are independent of the material properties and are, consequently, the same for the states of plane stress or plane strain. For $z=\pm a$, $\tau_{zy} = + p/\pi$ at y=0. This value remains approximately constant for a length of y=a and then drops parabolically until at approximately y=8a, its value becomes equal to 0.02p. The shear stress τ_{zy} at z=1.1a, on the other hand, is equal to zero at y=0, and increases rapidly to a maximum value of 0.29p at y=0.8. Beyond this value of y, its magnitude decreases gradually, and no significant difference in the shear stress values exists beyond y=2a for z=a and z=1.1a.

The normalized shear stress τ_{zy} in the collar joint of the composite wall, calculated using the transverse finite element model, is also plotted in Fig. 17. It is obvious that the half-space solution should not be compared with the finite element solution unless the boundary stresses at $z=\pm a$ are eliminated by some form of superposition principle. Nevertheless, the influence of these boundary stresses on the shear stress distribution near the point of application of the load should not be very significant. It can be seen that the shear stress predicted by the finite element solution is qualitatively very similar to that obtained by the closed-form solution of the half-space. The maximum shear stress values are within 70-80% of each other. Consequently, it may be concluded that the shear stress distribution in the collar joint calculated from the use of the transverse finite element model can be accepted to be correct with a good degree of confidence.

CONCLUDING REMARKS

From the analyses of the composite masonry walls subjected to in-plane loads on one wythe presented in the previous sections, one can conclude that the shear stress distribution in the collar joint is essentially nonuniform. The largest shear stress occurs near the point of load application which could be a possible weak spot for failure and crack initiation. It is very important, therefore, that appropriate crack initiation and failure criteria be established for the safe design of collar joints of composite masonry walls. These safe design loads can be derived through a combination of analytical and experimental research. In specific, experimental methodologies must be found through which shear strain variations in the collar joints can be measured.

It has been shown in this paper that the longitudinal model can be successfully used to compute longitudinal shear stresses in a collar joint due to discontinuous loads or wall openings. However, the magnitude of transverse shear stress in the collar joint (from one wythe to another) is overestimated by the use of the longitudinal model. It is recommended, therefore, that the plane strain transverse finite element model be used in general for computing transverse shear and normal stresses in the collar joint, even for cases where the applied load is discontinuous.

Lastly, it is recommended that load cases other than those considered in this paper must be investigated to get a better understanding of the collar joint behavior. In particular, horizontal in-plane loads on the brick or block wythes, that may act due to wind or earthquake phenomenon, must be considered in the analysis.

ACKNOWLEDGEMENTS

The research reported in this paper was supported by Grant No. CEE-802667 from the National Science Foundation. Computations were carried out on the IBM 370/3033 computer of Clemson University. The financial support of NSF and the cooperation of the Clemson University Computer Center are gratefully acknowledged.

REFERENCES

1. Anand, S.C., and Young, D.T., "A Finite Element Model to Predict Inter-Laminar Shearing Stresses in Composite Masonry," Proceedings, NSF Supported Conference on Research in Progress on Masonry Construction, ed. J.L. Noland and J.E. Amrhein, Marina Del Rey, CA, March 1980, pp. 5.1-5.28.

2. Anand, S.C., and Young, D.T., "A Finite element Model for Composite Masonry, Journal of the Structural Division, ASCE, Vol. 108, No. ST12, 1982, pp. 2637-2648.

3. Anand, S.C., Young, D.T., and Stevens, D.J., "A Model to Predict Shearing Stresses Between Wythes in Composite Masonry Walls due to Differential Movement," Proceedings, 2nd North American Masonry Conference, University of Maryland, College Park, MD, August 9-11, 1982, pp. 7.1-7.16.

4. Anand, S.C., and Young, D.T., "Load Transfer Mechanism and Resistance Capabilities of Composite Masonry Walls," Proceedings, 7th European Conference on Earthquake Engineering, Athens, Greece, Sept. 20-25, 1982, pp. 409-416.

5. Anand, S.C., and Gandhi, A., "A Finite Element Model to Compute Stresses in Composite Masonry Walls Due to Temperature, Moisture, and Creep," Proceedings, 3rd Canadian Masonry Symposium '83, University of Alberta, Edmonton, Alberta, June 6-8, 1983, pp. 34.1-34.20.

6. Anand, S.C., Stevens, D.J., and Brown, R.H., "Development and Application of the Finite Element Method to the Modelling of Composite Masonry Walls, Department of Civil Engineering, Clemson University, Clemson, SC, Report No. 4S-83, 1983.

7. Anand, S.C., Gandhi, A., and Brown, R.H., "Development of a Finite Element Model to Compute Stresses in Composite Masonry Due to Creep, Moisture and Temperature," Department of Civil Engineering, Clemson University, Clemson, SC, Report No. 5S-83, 1983.

8. Anand, S.C., McCarthy, J., and Brown, R.H., "An Experimental Study of the Shear Strength of the collar Joint in Grouted and Slushed Composite Masonry Walls," Department of Civil Engineering, Clemson University, Clemson, SC, Report No. 7S-83, 1983.

9. Anand, S.C., and Dandawate, B., "Creep Modelling for Composite Masonry Walls," Proceedings, 5th ASCE-EMD Specialty Conference, University of Wyoming, Laramie, Wyoming, Aug., 1-3, 1984, pp. 432-437.

10. Anand, S.C., and Stevens, D.J., "Computer-Aided Failure Analysis of Composite Concrete Block-Brick Masonry," Proceedings, International Conference on Computer-Aided Analysis and Design of Concrete Structures, Split, Yugoslavia, Sept. 17-21, 1984, pp. 649-661.

11. Anand, S.C., Dandawate, B., and Brown, R.H., "A Finite Element Model for Creep in Composite Masonry," Department of Civil Engineering, Clemson University, Clemson, SC, Report No. 20S-84, 1984.

12. Anand, S.C., and Stevens, D.J., "A Simple Model for Shear Cracking and Failure in Composite Masonry," Proceedings, 6th International Congress on Fracture, New Delhi, India, Dec. 4-10, 1984, pp. 2915-2922.

13. Anand, S.C., and Dandawate, B., "A Numerical Technique to Compute Creep Effects in Masonry Walls," Proceedings, 3rd North American Masonry Conference, University of Texas at Arlington, Arlington, TX, June 3-5, 1985, pp. 75.1-75.14.

14. Stevens, D.J., and Anand, S.C., "Shear Stresses in Composite Masonry Walls Using a 2-D Model," Ibid, pp. 41.1-41.15.

15. Brown, R.H., and Cousins, T.E., "Shear Strength of Slushed Composite Masonry Collar Joints," Proceedings, 3rd Canadian Masonry Symposium '83, University of Alberta, Edmonton, Alberta, June 6-8, 1983, pp. 38.1-38.16.

16. Cousins, T.E., Brown, R.H., and Anand, S.C., "Shear Strength of the Collar Joint in Composite Masonry Walls," Department of Civil Engineering, Clemson University, Clemson, SC, Report No. 1S-83, 1983.

17. Grimm, C.T., and Fowler, D.W., "Differential Movement in Composite Load-Bearing Masonry Walls," Journal of the Structural Division, ASCE, Vol. 105, No. ST7, 1979, pp. 1277-1279.

18. Porter, M., Ahmed, M., and Wolde-Tinsae, A., "Preliminary Work on Reinforced Composite Masonry Shear Walls," Proceedings, 3rd Canadian Masonry Symposium '83, University of Alberta, Edmonton, Alberta, June 6-8, 1983, pp. 15.1-15.12.

19. Redmond, T.B., and Patterson, D.C., "Compressive Strength of Composite Brick and Concrete Masonry Walls," Meeting Preprint No. 2515, ASCE National Structural Engineering Convention, New Orleans, LA, April 14-18, 1975.

20. Self, M.W., "Design Guidelines for Composite Clay Brick and Concrete Block Masonry. Part I - Composite Masonry Prisms," Department of Civil Engineering, University of Florida, Gainesville, FL, Research Report, April 1983.

21. Whitcomb, J.D., Raju, I.S., and Goree, J.G., "Reliability of the Finite Element Method for Calculating Free Edge Stresses in Composite Laminates," Computers and Structures, Vol. 15, No. 1, 1982, pp. 23-37.

22. Williams, R.T., and Geschwindner, L.F., "Shear Stress Across Collar Joints in Composite Masonry Walls," Proceedings, 2nd North American Masonry Conference, University of Maryland, College Park, MD, Aug. 9-11, 1982, pp. 8.1-8.17.

SUBJECT INDEX
Page number refers to first page of paper.

Brick masonry, 1, 87
Buckling, 71

Cavities, 46
Composite masonry, 106
Concrete slabs, 71
Crack propagation, 19

Eccentricity, 87

Finite element method, 1, 19
Floors, 71, 87

Incremental loading, 1
Interactions, 71, 87

Lateral loads, 46
Load bearing capacity, 71

Masonry, 19, 56

Progressive failure, 1

Shear strength, 56
Shear stress, 106
Shear walls, 56
Slabs, 87

Transverse shear, 19

Ultimate strength design, 56

Walls, 19, 46, 71, 87, 106

AUTHOR INDEX
Page number refers to first page of paper.

Anand, Subhash C., 106

Carruolo, Fred A., 46

Dhanasekar, Manicka, 1
Drysdale, Robert G., 19

Essawy, Ahmed S., 19

Ganz, Hans Rudolf, 56

Hamid, Ahmad A., 46
Hendry, A. W., 87

Kleeman, Peter W., 1

Leicher, E., 71

Mann, W., 71
Mirza, Farooque A., 19

Page, Adrian W., 1

Thürlimann, Bruno, 56